COLD MATTERS

# COLD MATTERS

*The State and Fate*
*of Canada's Fresh Water*

ROBERT WILLIAM SANDFORD

**RMB**
Victoria Vancouver Calgary

Rocky Mountain Books
www.rmbooks.com


Library and Archives Canada Cataloguing in Publication

Sandford, Robert W.
Cold matters : the state and fate of Canada's fresh water / Robert William Sandford.

Includes bibliographical references and index.
Also issued in electronic format.
ISBN 978-1-927330-20-3 (HTML).—ISBN 978-1-927330-46-3 (PDF).
ISBN 978-1-927330-19-7 (pbk.)

1. Ice—Canada. 2. Snow—Canada. 3. Climatic changes—Canada.
4. Freshwater ecology—Canada. I. Title.

QC981.8.I23S25 2012          551.310971          C2012-903848-2

Front cover photo: *Angel Glacier, Jasper National Park* © Wirepec


Printed in Canada

Rocky Mountain Books acknowledges the financial support for its publishing
program from the Government of Canada through the Canada Book Fund (CBF)
and the Canada Council for the Arts, and from the province of British Columbia
through the British Columbia Arts Council and the Book Publishing Tax Credit.

 Canadian Heritage  Patrimoine canadien  Canada Council for the Arts  Conseil des Arts du Canada

 BRITISH COLUMBIA ARTS COUNCIL
Supported by the Province of British Columbia


This book was produced using FSC®-certified, acid-free paper,
processed chlorine free and printed with vegetable-based inks.

*To the world's often beleaguered water and climate scientists, with hope for the future; and to Scott Wing, who is an expert on the relationship between fire and water.*

# CONTENTS

# PREFACE

It is with great pleasure that we write this preface to Robert Sandford's *Cold Matters*, on the science and implications of research funded by the Canadian Foundation for Climate and Atmospheric Sciences (CFCAS). The research was undertaken under two CFCAS networks that focused on how climate and humans affect ice, snow and runoff in Canada: the Improved Processes Parameterization & Prediction Network (IP3) and the Western Canadian Cryospheric Network (WC²N).

Over the last six years, we have become keenly aware of Bob's work in organizing meetings about climate change and water resources issues as well as his many popular books on these subjects and his eloquent public speaking about them. His work, and our respect for it, developed into a fruitful collaboration that allowed us to communicate our research findings to government, industry and communities. Bob has passionately supported our research and he has spent much time to develop a comprehensive understanding of the science, the effort behind it and its broader implications for society. When an opportunity arose for additional funding from CFCAS for outreach, we jointly proposed this book, with Bob as the intended author. He has translated and distilled the complexities of these scientific issues into an engaging book that we hope will have broad appeal to those who have an interest in snow, ice, climate and water. It is also hoped that the book will be well received by readers new to these topics.

Being part of IP3 and WC²N has been a wonderful experience and an honour for both of us. We have witnessed the excitement of new scientific discoveries and the pride in seeing students learn and blossom into the next generation of Canadian scientists. We have also been tremendously fortunate to have been supported by the Canadian Foundation for Climate and Atmospheric Sciences. CFCAS has been much more than a funding organization: it has been a valued partner in science and outreach. It is, in our opinion, a successful model for how science funding should operate. Canada was fortunate to have had the vision to establish CFCAS, and Canadians will benefit for many years from the advances in science that have occurred under CFCAS. IP3 and WC²N also benefited from support from the International Polar Year, the Natural Sciences & Engineering Council of Canada, BC Hydro, Columbia Basin Trust, Environment Canada, Natural Resources Canada, the Geological Survey of Canada and the governments of Alberta, British Columbia, Northwest Territories, Yukon, Saskatchewan, USA, Germany and UK, among others.

In 2005 the founders of IP3 suggested that in just over five years we could significantly improve our understanding of hydrological, snow, frozen ground and glacier processes in Canada's cold regions so as to improve the performance of hydrological, meteorological and climatological models of western and northern Canada. The IP3 founders further proposed to demonstrate improvements in hydrometeorological prediction using these methods.

At the same time, the founders of WC²N suggested that in five years WC²N could better understand the climate system and its effect on glaciers in British Columbia and Alberta through measurement of glacier extent over the past 400 years; documentation of regional climate variability; better understanding of cold regions processes and how they nourish glaciers; and modelling of changes in glacier cover and runoff for future climates.

Having a complex challenge like this is naturally exciting, even invigorating, and we feel that this ambitious plan has been realized through the tremendous hard work, energy and talent of our scientists, engineers, field technicians, students and managers. Both networks exceeded their expectations and produced a revolution in our understanding of cold regions water, ice and snow systems in Canada. This understanding has helped to produce a suite of improved glaciological, hydrological and land-surface hydrology models that are coupled to climate and weather prediction models and are being used to model glaciers and glacier meltwater in the Canadian Rockies, the Coast Mountains, the Columbia and South Saskatchewan river basins, the Great Lakes, and smaller basins in British Columbia, Alberta, Yukon, NWT, Saskatchewan and Ontario and in fact throughout the world, from Chile to China, Spain to Switzerland.

By linking with atmospheric models, the glacier and hydrology models are freed from some of the constraints – such as limited surface observations and changing climate – which were impeding our ability to predict glacier mass balance, snowpack, frozen ground, lake dynamics and streamflow. As with any science adventure we have ended up in places that we did not intend to go. IP3 was not fundamentally a study to observe climate change, yet all investigators found dramatic changes occurring in their research basins, with rapidly declining permafrost cover, snowpacks, glaciers and changes in the hydrological regimes, from less streamflow in the south to more flow in many parts of the north. The hydrometeorology of Canada's cold regions is changing as fast as we can understand it – this moving target doubled the challenge of IP3. WC$^2$N was not primarily a parameterization effort but some superb parameterizations developed from it that have improved climate model downscaling in mountains, mountain hydrometeorology and glacier dynamics descriptions in models. We constituted both networks with a regional focus but both have expanded to be relevant and linked in to many other parts of the world.

In short, the last six years has been a tremendously exciting adventure for members of our networks, from battling hypothermia on a remote mountain, to finding our way through blizzards on high glaciers or remote Arctic tundra, to travelling great distances over remote subarctic muskeg, high mountains and large lakes, to debugging complex computer programs, solving differential equations, fixing electronic instrumentation, cleaning up massive datasets, conceptualizing new and very abstract ideas as mathematical models, to communicating effectively in our 20 meetings. We have found all of this to be tremendous fun, and that lively energy has invigorated our science, the networks and our meetings and has given Canadian glaciology and hydrology a formidable reputation around the world. Through outreach we have tried to take new science to the people, governments and industry that can use it across western and northern Canada. We did more than we initially promised to CFCAS in all of this. Sure, we focused on the science, but we also provided a little entertainment at our workshops.

These networks have been about people as well as science. They have been networks of many dozens of interesting and talented colleagues from around the world and a dynamic training ground for over 100 students who are becoming scientists, engineers, technicians and educators in their own right. The human legacy of joint discovery, new collaborations and the ability to work on large problems will remain long after both networks cease to exist. We have benefited from devoted and skilled secretariats which gave us the support we needed to complete the science and which produced models and datasets as well as outreach, financial accountability and other reports that we can be proud of. Each network had a science committee and board of directors that were both demanding and inspiring and kept us on track.

This book is therefore a public legacy for these science networks that transcends the traditional channels of peer-reviewed journal papers, textbooks and ephemeral public talks or newspaper articles. We hope you enjoy it as much as we enjoyed the activities that are summarized in it.

—John Pomeroy, University of Saskatchewan
—Brian Menounos, University of Northern BC

Dr. Pomeroy and Dr. Menounos are the principal investigators and the science committee chairs of IP3 and WC$^2$N, respectively.

# INTRODUCTION

## WHY COLD MATTERS TO CANADIANS – AND TO EVERYONE ELSE IN THE NORTHERN HEMISPHERE

What winter does to water is the very essence of what the rest of the world imagines when it thinks of Canada. Much is made of snow and ice here. Though we sometimes curse the cold and the snow it produces, we accept and embrace both. We are a winter people. We shovel the snow off our sidewalks and then ski on the rest. After we scrape ice from our windshields, we drive to the rink to watch our kids skate on the stuff. We put up with the cold and we joke about it. Canada wouldn't be Canada and we wouldn't be Canadians if it weren't for the snow and the ice. Cold matters to Canadians.

Our annual triumph over the challenges of the Canadian winter is largely attributable to our success in adapting to the properties of water in its solid as well as its liquid form. But despite all the ways in which it influences our identity, few Canadians truly understand the amazingly complex processes through which liquid water becomes ice and snow. Fewer still appreciate the wide-ranging ecological and cultural implications of what ice and snow do to shape the physical landscapes that are the foundation of who we are as a people.

Because it surrounds them always, Canadians take water in its various forms for granted. Most of the world's fresh water is not available to us in a liquid form. Nearly 70 per cent of the world's fresh water exists in a frozen state in glaciers, perpetual snow and ice, and permafrost. This proportion of the planet's freshwater assets that remain locked up in solid form as ice is called the cryosphere. The cryosphere covers about 5.7 per cent of the earth's surface. It is composed of all of the world's sea ice and icebergs and all the ice in the planet's ice caps and glaciers. While presently accounting for only 2 per cent of the water that exists on the surface of the earth, it is the 2 per cent that means the most to us: the cryosphere contains nearly three-quarters of the fresh water that exists on the planet at any given time.

While the greatest concentrations of water in a frozen state are found in Antarctica and Greenland, significant icefields and glaciers exist in Canada. In fact, if we include all the water frozen in permanent snow, glaciers and permafrost, Canada possesses almost 20 per cent of all the fresh water on earth. In some ways, however, this is a highly deceiving statistic. The fact is that because this water is trapped as ice or is found in deep lakes left over from earlier, colder periods in our planet's history, it becomes available to us only once: when it melts. Water produced by the natural hydrological cycle, on the other hand, renews itself annually. When we subtract that water from our total supply we discover that Canada actually possesses only about 6.5 per cent of the water that circulates through the global hydrological cycle each year. As this book will evidence, however, the water stored in snow and ice in Canada is disappearing far faster than we expected. The amount of water made available to us each year through the hydrological cycle is also changing, as is the ratio of how much precipitation falls as rain instead of snow. While one might not think this matters, it does, as we will see.

Depending on where you live in Canada, up to half of the precipitation made available to you annually through the global hydrological cycle may arrive in winter in the form of snow. We are beginning to notice, however, that something is happening to winter that is affecting both snow and ice in

Canada. Winters are growing warmer and shorter. Patterns of snowfall and depth of deposition are changing. Whatever it is that is happening to winter has reminded us that cold matters to us, not just because it makes us hardy as a people but because of what it does to snow.

Snow cover acts as an energy bank which stores and releases energy over time. The absorption and release of energy throughout the year produces changes in the snowpack that are equivalent to seasonal changes in the landscapes upon which the snow collects. Once it is recognized that snow and ice are simply different forms of water, we see that most Canadians have as close an association with water as anyone who might live on a seacoast. Though we may not see it as such, the two-metre-deep snowdrift present for half a year outside your door might as well be a lake and the winter snowpack a freshwater sea rising and falling with an annual tide of seasonal cold. That snowpack is likely to be alive, too. Snow is a medium of life in its own right. It is habitat, providing both a place to live and a food supply for micro-organisms, invertebrates and small mammals. One study revealed that some 446 organisms are adapted to living in or under snow. Whole ecosystems exist in the blinding whiteness of the nival world.

Snow is also a medium of water transport. Snow is widely relocated by wind and intercepted by vegetation. When that snow melts it provides crucially important water to the ecosystems where it has collected. Perhaps most importantly, snow is also a reservoir that stores water in precisely the way a dam does for gradual release downstream over the course of the spring and, in many mountain ranges, long into the summer. There is not enough money in the world to build all the dams that would be required to store all the water that the winter snowpack does for later release into streams and rivers. Cold provides this invaluable service for free.

We are not the only ones who are suddenly realizing how much cold counts. In *Snow Ecology: An Interdisciplinary Examination of Snow-Covered Ecosystems*, edited by H.G. Jones, John Pomeroy, D.A. Walker and R.W. Hoham, we are offered a background introduction to the fundamental physical and chemical processes that characterize the evolution of snow and the feedback mechanisms between snow cover and the life forms and ecosystems of the world's cold regions. We learn that, on average, 23 per cent of the globe is covered at any given time by snow. The effect of snow cover is most pronounced where the global land mass is concentrated. Mean monthly air temperature can drop by 5°C just due to the presence of snow. Because it normally covers more than half of the land in the northern hemisphere each year, and possesses such important properties, seasonal snow cover is recognized as a defining ecological factor throughout the circumpolar world.

One of the qualities that makes snow such a central influence is its amazing insulation properties. The temperature of the upper surface of a 10-centimetre-deep snowdrift may drop by more than 10°C overnight, but the temperature of the underlying ground may drop by only a single degree. By keeping the near surface of the earth from cooling to the same temperature as the atmosphere, snow cover keeps the roots of trees and other plants relatively warm, preventing damage from freezing. And it is not just ecosystems that benefit from this effect. We benefit too. In many parts of Canada the presence of deep snow is what keeps the soil warm enough to grow grain crops like winter wheat. But the effects of snow cover do not stop with its warming influence on surface soil. Snow's effect on ecosystems is also huge when it melts.

Snow also functions as a radiation shield that reflects most shortwave radiation but absorbs and re-emits most longwave radiation. In reflecting light, snow also redistributes heat. The slow retreat of regional seasonal snow cover ensures that the ground remains cold and saturated. This delays the increase in temperature in springtime in western Europe, for example, for weeks. Similar effects are clear in Russia and Canada.

Because it also increases the amount of radiation lost near the earth's surface, the presence of snow affects climate in ways that are remote from the snow cover itself. It appears that the presence of seasonal snow cover plays such an important role in the global climate system that it can be argued that all ecosystems – everywhere – are indirectly affected by snow cover.

Snow cover, atmospheric circulation and temperature are interdependent and relate to one another as feedbacks. Temperature to a large extent defines how, in what form and how quickly water moves through the global hydrological cycle. Water and temperature define climate. Climate defines ecosystems. And ecosystems define us. In the absence of snow we would be different people living in a different world. It appears that cold really does matter.

This book explains how and why cold matters to Canadians and ultimately is important to the health and well-being of everyone on earth. It tells the story of how the workings of science allowed two networks of Canadian researchers to confirm the extent to which our changing global climate has undermined the stability of our relationship to cold as expressed by what cold does to water in the form of snow and ice. This book aims to demonstrate how collective scientific breakthroughs are allowing us to visualize how temperature affects hundreds if not thousands of interdependent natural processes related to the temperature and state of water, not just in Canada but throughout the circumpolar world. Ultimately, this book is about discovery, for it tells the story of how two research groups collectively composed of nearly a hundred scientists and graduate students together confirmed that the hydrological cycle in Canada is changing.

The quiet scientific revolution these researchers have started had its genesis in the realization that temperature is a fundamental parameter in the eternal interaction linking water, life, weather and climate that makes existence on this planet not only possible but meaningful in human terms. What these researchers have discovered is that cold matters a lot more than we thought in the context of where and how we live in Canada. Because of the collective efforts of these researchers we are beginning to know precisely why cold matters so much.

Finally, this book is about how science is struggling, in the face of many unnecessary obstacles, to create new branches of mathematics and new statistical analyses that will enable us to more completely understand, model and predict how upward changes in temperature will further affect how water moves through the hydrological cycle in Canada and elsewhere in the circumpolar North. This is not a process that is going to happen overnight, however. Canada has fallen 30 years behind in research and understanding of the hydrological processes that are taking place, especially in the Arctic and in the cold regions of its many mountain chains. This matters because such research is essential for improved ecosystem management, water management and water security as it relates to ensuring sustainable supply now and in the future.

There is no reason, at the moment at least, why this story can't have a happy ending. The breakthroughs this book describes point in the direction of possibility and hope. The work of these two research networks makes it apparent that what happens to water is the clearest and most visibly obvious expression of what climate change may mean to us as a society in the future. If we can more fully characterize all the extraordinary ways in which water in its different forms together influence geophysical processes and natural ecosystem function, we will be better able to understand how these collective processes cumulatively influence the weather, which over the longer term defines climate. Once we are able to do this, it will become possible to more clearly identify what habits we have to change in order to sustain our civilization within the ecological and climatic boundaries that circumscribe practical adaptive possibility.

Bob Sandford
Canmore, Alberta
June 2012

# CHAPTER 1

## The People, the Places, the Prize: Understanding the Present, Predicting the Future

It is mid-January on the Canadian prairies, but something is wrong. There is no snow on the ground and the temperature is +12°C. It is so warm that no one is wearing a winter coat. It is warm enough that when we drove into the small prairie city of Red Deer, Alberta, where a hydrological modelling course was to be offered, we saw a man driving a convertible with the top down.

One of the reasons many Canadians are ambivalent about climate change is that they don't see how a slight rise in temperature could hurt, especially during our long, frigid winters. Environment Canada's Fred Wrona spoke for a lot of Canadians when he joked at a major climate conference in 2004 that Canada wasn't getting warmer, just less cold. In the context of what temperature does to water, however, models that mimic natural hydrological processes invariably demonstrate that cold matters. In fact, it *really* matters.

Hydrological modelling courses attract unlikely audiences. Because you need to know a lot about water to be able to predict what it might do on the landscape under any given set of conditions and temperatures, the people attracted to these courses are usually comfortable dealing with higher mathematics. They also tend to be people who are responsible to the public for ensuring adequate water supply or for preventing too much water from causing flood damage to houses, roads and bridges. These courses also tend to attract people who are very concerned about what climate change might do in terms of future water supply to natural ecosystems. But hydrological forecasting does not concern itself only with ecosystems or climate. Short-term forecasting and long-term prediction of hydrological change is a highly practical matter. Knowing how much water is going to be available is a make-or-break proposition for Canadian agriculture. Knowing how much water there is going to be when and where in the future is crucial to understanding long-term water supply availability in town and cities. Without this knowledge it will be difficult to determine how big a city or a region can become in terms of population growth or whether or not irrigation agriculture will work in a given region and to what extent.

Understanding how much water will be available in the future is also central to flood prediction and control, the design of water and sewer systems, the location of roads and bridges and the standards to which they should be constructed, and the planning of subdivisions, especially if they are located near highly desirable river or lake views. Knowing what will happen with water security over the long term also helps governments ensure they can honour existing water licences to cities, industry, agriculture and their downstream neighbours without sacrificing the health of aquatic ecosystems, contaminating surface or groundwater or generating unanticipated human health hazards.

It is impossible to plan wisely for the future if you do not know how much water you will have and what its quality will be under a variety of projected scenarios. Modern planning favours the modelling of these scenarios so that we don't learn by accident that we have miscalculated our effects on landscapes and our influence on the atmosphere, or that we have underestimated the combined or cumulative influence of those factors on how much water the hydrological cycle produces in any given place at any given time. We know clearly from history the consequences of such mistakes and we do not want to repeat them. As all of these impacts appear to be accelerating, hydroclimatic modelling is being taken a lot more seriously than ever before.

Climate models, as Heidi Cullen explains in her 2010 book *The Weather of the Future: Heat Waves, Extreme Storms and Other Scenes from a Climate-Changed Planet,* are built from two types of equations. There are physics-based simulations that rely on elegant equations such as Newton's laws of motion and conservation of energy. Then there are simulations based on what are known as parameterizations – equations that are derived from direct observations and that attempt to represent our current understanding of how our climate works to generate local weather conditions.

The physics used in contemporary modelling is universal, whereas parameterizations will vary depending upon local conditions modellers want to represent in their model. Parameterizations are a way to estimate the vast array of complex interactions among land, life, air, water and weather that are observed in nature but whose physics are so complicated that they can't be reproduced exactly in models due to limitations in computing power and speed. Each climate model uses different parameterizations to approximate what it cannot represent directly. As a result, different climate models project different degrees of future warming.

No matter which model you use, parameterizations inevitably introduce uncertainty. This is why major climate assessments, such as those conducted by the Intergovernmental

Panel on Climate Change, draw upon the results of up to 20 different models which together constitute an ensemble of model simulations. This ensemble approach utilizes the results of multiple models to reach a consensus on likely future scenarios. Weather forecasters also use this technique to improve their predictions. The assumption behind this approach is that approximation errors among models will tend to cancel each other out when the projections are averaged, leaving the most robust tendencies intact.

As Cullen points out, climate models have now reached a level of sophistication that approaches if not rivals that of weather models. Current models can parameterize grid areas of about 110 square kilometres while incorporating data from 26 vertical layers representing differing and ever-changing conditions from ground level to various strata in the atmosphere. The next generation of models will be capable of resolving down to a grid resolution of about 50 square kilometres. As computers become faster and more powerful, it is not impossible that models may one day be able to predict the climate for every square kilometre on our planet. But that prize has yet to be won.

## THE ROOTS OF HYDROLOGICAL MODELLING IN CANADA

Dr. John Pomeroy began his Red Deer presentation on the art of hydrological modelling by chronicling the development of the cold regions hydrological model, whose function he would be demonstrating over the course of the next two days. He explained that the first generation of global climate models, which began to appear in 1969, dealt with snow cover in a very simple way that distorted variations in snow cover over time and space. These early generations of models reproduced ground surface temperature under the snow with a margin of error of up to plus or minus 10 °C, rendering them of limited applicability for assessing the effects of snow cover on terrestrial ecosystems.

More advanced processes for the modelling of snow cover were eventually developed that allowed researchers to consider snow as a separate thermal layer from the underlying soil. Researchers like Environment Canada's Diana Verseghy were at the forefront of being able to simulate multilayer snow cover and to model heat transfer within the snowpack in a way that permitted the underlying surface to be treated separately.

## DIANA VERSEGHY

Diana Verseghy, of the climate research division of Environment Canada, has undertaken research into land-surface modelling in northern Quebec as part of Canada's involvement in the International Polar Year. This work involved improvements in parameterization of the Canadian Land Surface Scheme (CLASS), based on satellite imagery over vast areas of the Quebec Arctic. CLASS is a key component of the MESH hydrological model and is the land component of climate models in Canada. The use of satellite-based remote sensing has the potential to increase the reliability of climate and hydrological models in Canada in the future. With imagination, hard work and further support, Canada's hydrological science community can lead us to sustainability.

While not as sophisticated as today's models in terms of simulating snow cover, even the earliest iterations of models such as those developed by the Canadian Climate Centre suggested that enhanced warming was likely in those regions that are subject to seasonal snow cover in Canada. There was concern, because of the almost instantaneous feedbacks that occur between snow cover and the atmosphere, that terrestrial ecosystems presently affected by variations in spring snow cover could be put under considerable pressure as a result of continuing climate change. In addition to the reduced seasonal snow cover brought about by warming temperatures induced by a doubling of the $CO_2$ equivalent in the atmosphere, researchers projected a general increase in soil moisture content in northern latitudes because of greater precipitation, faster snow melt and a small increase in the rate of potential evaporation due to warmer winter temperatures. The net impact of all these changes was that the prairies should expect to be drier in the summer.

The same models, however, also projected that over time there would ultimately be a significant decrease in snow cover and snowpack throughout the year. One climate experiment using the Canadian Climate Centre $2 \times CO_2$ model projected a 20 per cent reduction in winter snow cover in the northern hemisphere and as much as a 50 per cent loss in snow cover that lingered through summer. Other models projected that what was a plausible $4\,°C$ warming at the centre of the North American continent could result in a 70 per cent decrease in snow cover duration over the larger Great Plains region and a 40 per cent decrease over the Canadian prairies. The models that were used to make these projections had proved their worth. The next step was to develop models that would more accurately predict hydrological changes that might take place under a variety of scenarios at much smaller, local scales.

Pomeroy pointed out that the development of the cold regions hydrological model began in the 1990s with a land-use model created at the National Water Research Institute in Saskatoon. This model was created by a much younger

Pomeroy himself, along with his now-deceased mentor – and founder of Canadian hydrology – Don Gray and now-retired engineering genius Tom Brown. Pomeroy went on to enumerate the changes and advancements in the model over the last decade, which clearly demonstrated that through constant testing of parameters and improvement in processes, modelling really did result in breakthroughs in knowledge and predictive capacity.

## THE COLD REGIONS MODEL AND CFCAS: A DECADE OF ADVANCEMENT IN CANADIAN CLIMATE SCIENCE

Pomeroy explained to his Red Deer conference audience that the cold regions hydrological model, or CRHM (pronounced "crim"), emerged in its current iteration as a result of investments made in research networks by the Canadian Foundation for Climate and Atmospheric Sciences. The foundation, or CFCAS as it was known, was created in 2000 in response to the government of Canada's commitment to atmospheric and related ocean and water research as it pertains to the environmental, social and economic future of the country. In February of that year, the federal government earmarked $60-million over six years to create an autonomous foundation whose goal would be to "enhance Canada's scientific capacity by funding the generation and dissemination of knowledge in areas of national importance and policy relevance, through focused support for excellent university-based research in climate and atmospheric sciences." The work was later deemed so important that the CFCAS received a further $50-million from Ottawa in 2003. Its mandate was also extended until March of 2011.

The linked networks that came into existence as a result of CFCAS over the decade it was in existence include the Drought Research Initiative, which studied the causes and implications of the 2001 to 2004 drought on the Canadian prairies; the Western Canadian Cryospheric Network, or wc²n, which studied climate change effects on Canada's glaciers; and the Improved Processes Parameterization & Prediction Network, or ip3, which sought to improve hydrological and atmospheric prediction at regional and smaller scales in the cold regions of Canada.

ip3 devoted its efforts to understanding water supply and weather systems in cold regions at high altitudes and high latitudes in Canada. While by definition ip3 focused principally on the Rockies and western Arctic, the outcomes of the research it conducted were of great interest in other parts of the country as well, if only because there is no place in Canada where cold doesn't somehow matter in terms of water, weather and climate.

The aim of ip3, which ran from 2006 until its funding ended in 2010, was to contribute to better prediction of regional and local weather, climate and water resources in cold regions, including those river basins in which streamflow is not presently gauged. ip3 also explored how changes in snow cover will affect water supplies, with particular reference to freshwater inputs to the Arctic Ocean. The ip3 network comprised over 40 investigators and collaborators from Canada, the United States, the United Kingdom and Germany. As well as being a lead investigator in his own research projects, Dr. John Pomeroy was ip3's chair and also co-chair of its sister network, the Drought Research Initiative.

Though the improvement of the cold regions hydrological model was just one element of the research being conducted by the ip3 network, the model was already generating immediately useful and widely applicable results. At the time Pomeroy gave his presentation in Red Deer, the cold regions hydrological model was the first object-oriented, modular, physically based hydrological model in Canada. It had been adopted by two research networks and introduced in courses to over 100 students from across the country. The model was also being evaluated by the governments of Canada, Alberta, Saskatchewan, Manitoba, Yukon and Ontario for use in water resources assessments. The development of the cold regions hydrological model, Pomeroy explained, was a practical outcome of IP3 research that is now – like many other CFCAS products – available to any government department, agency, institution or private-sector interest that wishes to better understand how hydrological cycles work in the cold regions of Canada. But even Pomeroy thought the model could be improved upon.

Pomeroy is a scientist's scientist. He is cautious and constantly critical, even of his own findings. As one participant in the workshop later commented, Pomeroy's summary of the history of the cold regions hydrological model was an excellent example of just how science itself advances. It may be a plodding process, but the gradual, persistent improvement of ideas and techniques has proven itself time and again to be a powerful mechanism for the advancement of human knowledge. If indeed we are in a race between population growth and the accumulation and application of this kind of knowledge, our only hope is that this great method of scientific validation of reality can somehow keep up with the kinds of problems we are creating for ourselves through the explosion in our population and our material expectations. What researchers like John Pomeroy are doing is not just curiosity-based science. It is science that matters a great deal to all of us. Right now we can only guess at the extent of the damage we are causing to the world. Better models will help us understand what we can do to prevent that damage. Even more importantly, they may be able to direct us to specific action that will prevent our rapidly converging problems from robbing us of our prosperity and our future. Models of that kind will be worth having.

But as Pomeroy pointed out, models are not reality. Though they may be the best tools we have – and they will undoubtedly continue to get better – they still do not emulate the complexity of nature in all circumstances. The ultimate prize, Pomeroy believes, is to produce models that do just that. It is his aim – and the aim of many other IP3 researchers – to soon be able to mathematically simulate enough of the central eco-hydrological and hydroclimatic processes that take place in Western Canada to be able to accurately and reliably forecast what the hydrological cycle might do in any given year, and to project these forecasts accurately enough to be able to predict climate-change effects on the region over time. This prize is huge.

The benefits to agriculture alone would be enormous. If forecasts of water supply were more accurate, tied to improved seasonal weather forecasts and available earlier, farmers would be able to choose the most suitable crops for projected precipitation and other climatic patterns. It would become possible to more accurately predict droughts and floods earlier. We would also have a much better idea of what the "new normal" in terms of long-term climate might be in Canada's prairie and mountain West and throughout the Pacific Northwest.

The development of such predictive capacity will not be easy. It will require ever more complex models and huge computing power. As Pomeroy noted, there are many obstacles to overcome. Models, he explained, are mathematical abstractions of the essential elements of nature that allow some degree of prediction of outcomes of interest. Pomeroy observed that universal models are seldom accurate, principally because the scale explored in such models is often so large that the predictions that are derived from them are of little use locally. It is for this reason that many IP3 researchers were working to downscale their models to make their results more relevant where they live. It is also important to understand that the hydrological cycle is highly variable in different parts of the world. To date, no single modelling approach has been found to be applicable in all environments at all scales or to predict everything we may want to know. This local and regional variability has made it necessary to create hundreds of different models, each one capable of characterizing local aspects of the water cycle. Hydrological models, Pomeroy explained, seem to work best close to home, where local circumstances are more clearly understood.

The IP3 research network committed itself to bridging the gap between the real world and modelling. The difference between observation and simulation is held to be one of the most important gulfs to be bridged in contemporary science, especially in the domains of water security and climate change prediction. Without bridging this gap we will not be able to take what we are observing through direct monitoring in the field and translate it into useful predictions of what might be happening in natural systems in a variety of scenarios in the future. Without the harmonization of modelling results with what natural systems actually do over time, the predictions science makes about the future will not create a solid foundation upon which public policy decisions can reliably be made about matters such as water security, infrastructure design and climate adaptation. There are a lot of parameters to take into account.

## WHAT IS PARAMETERIZATION?

Parameters, as Heidi Cullen noted, are the distinguishing or defining characteristics or features of a natural system. They can be viewed as constant elements or aspects that serve as boundaries beyond which certain functions cease to operate. Parameterization means quantifying these key natural variables and adjusting them to suit a mathematical model so that the model can use them to more accurately represent nature in prediction simulations. Parameterization links the "perfect" natural environment with "imperfect" computer models of the environment, and knowing just which parameters to measure in a natural system is essential to understanding the dynamics of that system. Parameters and parameterization matter because improvements in measurement and choices of parameters to be measured are the way in which simulations in models are made more consistent with actual on-the-ground observation. Only by making simulation consistent with on-the-ground observation can we be sure that the right data is being collected at an accurate enough standard to enable it to become a reliable foundation for accurate and useful prediction that political and business leaders can rely on to make sound decisions in matters related to water and climate security.

As John Pomeroy will quickly point out, a parameter can also be viewed as a quality that is constant in the case considered but which varies in different cases. The fact that constants vary in different cases is what makes it necessary to seek different parameters in cold regions than one might in warmer places. The fact that some constants are unique to cold regions is what makes it necessary to construct models to interpret what is happening to our water in Canada that are very different from those that would be used in other, warmer parts of the world. There are a great number of distinctive hydrological elements that complicate modelling in the cold regions of Canada. These include snow storage; snow redistribution and melt; infiltration into frozen ground; how water moves through thick organic soils; the extent to which cool surfaces slow evaporation; the effects of frozen rivers and lakes; seasonal differences in energy inputs; and an increasingly sparse data network that makes it difficult for modellers to know as much as they would like about what is actually happening in nature so they can simulate processes accurately. Another unique problem we have in Canada is the difficulty of determining the actual drainage area in recently deglaciated basins where erosion has yet to establish clear drainage patterns.

The challenge in Canada, Pomeroy explained, is to choose a model that is consistent with the assumptions upon which you may wish to build. This means careful selection of parameters that influence the hydrological processes you want to model, so that resulting predictions mirror what is actually happening in nature. In cold regions this means that modelling parameters must at a minimum include snowmelt dynamics, the influence of permafrost, how much snow is intercepted by forest canopies, and differential patterns of evaporation from bodies of cold water, frozen ground or glaciated drainage basins. These parameters, Pomeroy insisted, have to be considered in addition to all the other, usual ones such as precipitation, runoff, evaporation, cloud formation, atmospheric advection and dozens of other energy and standard hydrological parameters. In other words, it is not easy to reproduce what happens all around us every day in nature. The pleasant world we cast our eyes upon each morning is almost unimaginably complex. It arises out of the interaction of an almost infinite number of relationships. Each of these relationships can be expressed as a parameter.

## BUILDING A MODEL THAT NEVER STOPS LEARNING

Because the cold regions hydrological model is a platform upon which it is possible to build more than just one model within

## HOW THE COLD REGIONS HYDROLOGICAL MODEL DIFFERS FROM OTHER MODELS

As hydrological modelling began to develop in Canada, it soon became apparent that models developed for application in temperate regions did not work in western and northern Canada, because they presumed drainage patterns that didn't exist in Canadian cold regions. The CRHM also came into existence as a response to the cumbersome nature of earlier models built on FORTRAN and other outdated computer languages based on code that stood in the way of improvements in graphical representation.

The cold regions hydrological model is a physically based system in that it depends upon field information to identify the principles governing the primary physical processes responsible for most water movement in a given basin. Field measurements also help establish the fundamental boundary and initial conditions that affect these processes, and provide a scale on which to measure the validity of these processes.

Because the CRHM is physically based, the results it generates contribute to a better understanding of basin hydrology and result in the characterization of process parameters that can be transferred to other cold regions. The CRHM hydrological cycle simulation is sensitive to the impacts of land use and climate change. It reflects the effects of landscape sequencing in natural drainage basins and does not require the presence of a stream in each land unit as other models do. The CRHM platform is also very flexible in that it can be compiled in various forms for specific needs and is suitable for testing individual hydrological process algorithms or other sets of rules used for computer based problem solving.

Another quality of the cold regions hydrological model is that it is easy for hydrologists to use and very useful for teaching. Its most important quality, however, is that, in the absence of monitoring in many regions of the country, CRHM applications do not need to be calibrated based on information from streamflow gauges and other monitoring outputs. Because monitoring has become so sparse in so many parts of Canada as a result of budget cuts by environment ministries, John Pomeroy and his IP3 colleagues realized they had to develop a hydrological cycle simulation that didn't require the calibration of streamflow measurements at the outset. While they have succeeded, this does not excuse governments from cutting back on monitoring, for without it even the best models will eventually fail for want of accurate data.

itself, it allows modellers to create simulations of what actually occurs in nature that are specific to local situations. The CRHM platform allows synthesis of data using mathematical functions and interpolating data. With this platform it is possible to change parameters such as forest cover in order to model, for example, the effects of deforestation or the hydrological implications of warming seasonal or mean temperatures. It is also possible to use CRHM to run several internal models in parallel so as to determine which parameters in which models are the most accurate in terms of their ability to represent what actually happens in nature. Errors in simulation can be used to identify gaps in understanding of processes, structure or parameters. Constant reiteration improves the model over time, making it ever more faithful to reality and ever more useful as a tool for long term forecasting and prediction. In this regard, this process can be characterized as being heuristic, in that it proceeds to solution through trial and error.

Viewed in this light, the cold regions hydrological model is not only a useful tool for predicting future hydrological outcomes in many parts of Canada, but also a valuable introduction to just how complex nature really is, in that the model demonstrates how one level of complexity in natural systems is embedded in the next. In only a few short hours of exploring the model it becomes abundantly clear how important water is to nature. Water in one or another of its phases is implicit in every natural life function and in nearly every physical process that defines our earthly reality. Moreover, it appears these processes are nested, with water performing different important functions at each level. Different parameters therefore exist at each level, nested one within another, with water complementing itself and reiterating its own diverse function throughout a nature it completely permeates. Reality is saturated with water and life made richly manifold by all the fundamental ways in which water reacts with almost every element in the physical world. Even the best models, however, reveal only a glimpse of how implicit water is in the construction and maintenance of our earthly reality over time.

Metaphysical and purely mathematical considerations aside, the cold regions hydrological model depends on clearsighted understanding of physical processes for its veracity. A careful analysis of parameters, Pomeroy explained, can produce model results that are useful in a broad range of practical situations. Changes, for example, in streamflow, soil moisture, wetland storage and groundwater dynamics in a given place can be modelled to demonstrate how sustainable the withdrawal of well water will be under a variety of future scenarios.

An even more comprehensive understanding of physical processes is necessary in the modelling and predicting of extreme situations such as floods and droughts. The understanding of physical hydrological processes thrives also on interdisciplinary understanding of chemistry and ecohydrology, which makes modelling more interesting and satisfying scientifically. These relationships in themselves elevate hydrological practice to hydrological science.

Models are improved by physical observations at multiple scales over time, requiring the creation of a wide variety of research basins which provide a range of models from which to choose parameters that will be relevant to outcomes a given researcher may wish to predict. The basic unit forming the foundation of any contemporary model, including the one developed by IP3 to better understand Canada's cold regions, is the hydrological response unit, or HRU. An HRU can be any segment of a basin that has uniform hydrological elements, such as a cultivated field, a pasture, a wetland or a lake.

Any given hydrological response unit will have three groups of attributes:

- a biophysical structure comprised, for example, of soils, vegetation, elevation, slope and area, data for which can be derived from maps or from geographical information systems;
- a definable hydrological state, which will be characterized by such parameters as snow, snow-water equivalent, internal energy, intercepted snow load, soil moisture and water table; and
- the particular HRU's own unique parameters of hydrological flux, or ways in which water moves around within the unit. In most Canadian contexts the way water moves about is determined by snow transport, sublimation, evaporation, melt discharge, infiltration, drainage and runoff.

In all of these processes, cold matters.

In building a hydrological model, researchers measure all three of these groups of attributes in any given hydrological response unit and they work outward to do the same in each adjacent HRU. The object is to connect the HRUs in a way that allows accurate characterization of outflows.

Constant improvement of the understanding and accurate depiction of the widest range of parameters is required to ensure increasing reliability of these models. Many of the researchers funded by the Canadian Foundation for Climate and Atmospheric Science through the IP3 network focused independently on specific parameterizations. Dr. Sean Carey of Carleton University, for example, is working to improve parameterizations for organic-covered permafrost soil in land surface and hydrological models. Carey found that accurate parameterization of organic soils and freeze–thaw dynamics in soil were major headaches in the development and utilization of both land-surface and hydrological models. Accurate water and energy simulations were difficult to create because of the wide range that existed in algorithms and parameterizations in these models. These problems were compounded by the fact that very little evaluation and validation of these parameters had actually been undertaken in the field in cold regions in Canada or elsewhere.

Carey and his team compared and evaluated algorithms and parameterizations in use in a number of current land surface and hydrological models at a number of test sites. They concluded that because empirical and semi-empirical algorithms required site-specific parameter calibration, they were not suitable for use in land surface and hydrological models designed to work across various site conditions. They determined that segmented linear functions were easiest to use for unfrozen water parameterization. They also determined that coupled numerical simulations were the best choice for characterizing freezing-point depression for water in simulation of soil heat and moisture transfers.

Translated into language the average person might relate to, Carey and his colleagues advanced common understanding of which algorithms work and don't work in characterizing what happens to water in soil and in freezing and thawing ground under various model simulations. Their findings will point others working in this important field in the right direction with respect to efforts to make simulation consistent with what we observe to happen in nature. One day we may actually know how the world beneath our feet actually works, beginning with what water does in its liquid and solid forms.

But Sean Carey is not alone in his efforts to address the challenges associated with parameterizations of complex interactions between soil and water. Dr. Ric Soulis, a brilliant mathematician on the civil and environmental engineering faculty at the University of Waterloo, is also working on the accurate parameterization of soil water in modelling processes. Soulis has noted that a simple soil parameterization scheme called Watdrain had been the standard for characterizing soil moisture in shallow aquifers. He is now proposing a different standard, based on a more rigorous application of an algorithm called the Richards equation. Dr. Soulis has mathematically demonstrated the procedures for applying this new parameterization in the context of the cold regions hydrological model and in other distributed hydrological land surface models. Other IP3 researchers, such as Sean Carey, are taking notice.

Cécile Ménard, then at the School of Geosciences at the University of Edinburgh, was also working on the further development of specific parameters. Ménard, in association with the principal investigator of her research team, Dr. Richard Essery, is working on snow processes and parameterization in complex landscapes. This work has been advanced through animations Ménard and her colleague developed to show how shrubs capture snow and how this capture later affects snowmelt.

Ménard has pointed out that in order to build higher-resolution models, it is necessary to make high-resolution characterizations of the landscapes you want to model. To this end, she and her colleagues were using LiDAR (Light Detection and Radar), a laser-based reflected-wave technology that is similar to the older, radio-based RADAR (Radio Detection and Ranging) but which can be used from an aircraft to measure the height of terrain and vegetation to an accuracy of just a few centimetres.

Next, this research team wants to investigate how reliably the parameters they have developed can be transferred to different sites. They also want to continue to improve the accuracy of their canopy parameterizations over larger areas of complex topography.

Once again the goal is to arrive at more accurate prediction of how hydrological processes will be affected by human impacts and climate change. At the moment, this work is barely keeping up with such changes. But it may nevertheless make it possible in the future to understand our impacts before we modify landscapes or the composition of our atmosphere instead of simply blundering ahead in the hope that no catastrophic effect will result.

Other researchers in the IP3 group were seeking ways to harmonize parameterizations from other models with the cold regions hydrological model. Working with John Pomeroy and Alain Pietroniro of the Water Survey of Canada, Dr. Edgar Herrera of the Centre for Hydrology at the University of Saskatchewan examined outputs from a model called Modélisation environnementale communautaire, or MEC, and from the Canadian Global Environmental Model, known as GEM. Dr. Herrera then integrated these outputs into the cold regions hydrological model to study snow transport over Marmot Creek in the Kananaskis region of the Alberta Rockies. This ongoing research is beginning to delineate important relationships between climate conditions and the persistence of snowpack and snow cover, which are crucial missing elements in the determination of climate impacts on future water supply throughout the western half of North America.

## ALAIN PIETRONIRO

Dr. Alain Pietroniro is the director of the Water Survey of Canada. He works with a number of land-surface hydrological models upon which the government of Canada relies for environmental prediction. His current focus is on the Modélisation environnementale communautaire, known among professional modellers as MEC, which was used to integrate upper-air observations of precipitation with land data. Another model he uses is MESH, a model linked to MEC which utilizes a surface and hydrology configuration designed to aid in regional hydrology modelling. MESH itself is based on CLASS, developed by Dr. Diana Verseghy, and WATFLOOD, developed by Dr. Ric Soulis. Pietroniro has demonstrated the success of a modelling methodology that includes parameters such as landscape topography and vegetation, snow accumulation regimes, blowing snow transport, snow energetics, and snow interception as well as runoff generation and the feedbacks which runoff creates.

Another IP3 collaborator, Russian hydrologist Olga Semenova, is working to create a universal hydrological model called Hydrograph. The first pronouncement she put forward concerned the physical scale of the research basins she worked to model in her homeland. The Lena River, Semenova explained, drains 2.5 million square kilometres, roughly the area of the entire North American Great Plains. The Lena is larger by far than even Canada's biggest river basin, the Mackenzie at 1.8 million square kilometres, and twice the size of the St. Lawrence, which drains just over a million square kilometres. Even mid-scale basins in Russia are enormous, defined, as Semenova explained, as a system of less than 200,000 square kilometres. At 140,000 square kilometres, the South Saskatchewan is a mid-sized to small river by Russian standards.

A second telling point made by Semenova was to demonstrate weaknesses in a number of conventional algorithms which she hinted were used widely in North American models. She showed how evolving Russian hydrological models in fact harmonize simulation and observation, not just for streamflow but also for soil moisture. She also demonstrated that the Russian model worked equally well in Angola and Costa Rica. Her evidence: an almost perfect correlation between actual observations and model predictions.

The obvious effect of Dr. Semenova's presentation was to cause the modellers in the room to continue to critically question the appropriateness of their parameterizations. This, of course, demonstrates once again the value of the scientific method and the importance of international collaboration on the ultimate meaning and value of all research outcomes.

In concluding her presentation, Dr. Semenova cautioned that accurate prediction ultimately depended on not being stuck with one model or with established approaches. The key to accurate predictive modelling, in her estimation, was the multiple testing of approaches over basins of all kinds, regardless of their scale, topography and climates.

## HOW TO HANDLE ALL THAT DATA

Parameterization is only one element in a suite of challenges that must be overcome if the forecasting and prediction of hydrological change is to become more accurate. In order to understand and simulate natural physical processes, large volumes of data describing a wide range of hydrometeorological phenomena had to be gathered first. In IP3 these typically included an enormous number of readings of temperature, pressure, humidity, wind speed and direction, incoming and outgoing radiation at long and short wavelengths, precipitation and snow depth, soil moisture and streamflow, all measured over several years and at anything from ten-minute to daily intervals in a variety of locations. For WC$^2$N, there were tree-ring and ice-core transects, associated inferences of past climatic conditions, remotely sensed imagery from satellites and aircraft, spatial coverages describing glacial extents, digital elevation models and hypsometry and ultimately huge four-dimensional grids describing time-variant conditions within cells at specific locations and elevations across entire mountain ranges.

Generating datasets of this type is always expensive: buying and maintaining complex instrumentation – and mounting field expeditions, often to remote areas, to gather the data – demands considerable outlay of both money and time. Both networks, with strong encouragement from the funding body, the Canadian Foundation for Climate and Atmospheric Sciences, felt it would be important to make the data gathered by their participants available to the broader scientific community and the general public as permanent legacies of their work. Michael Allchin, as data manager for IP3 and WC$^2$N, was tasked with establishing these archives.

In the past, researchers have often been reluctant to publish the full data on which their research hypotheses were founded, and Allchin sees a need for a broad cultural change toward more open data management. This change will not always be accepted willingly: in recent years, there have been signs that shady organizations and individuals exist who are willing to adopt nefarious means to discredit environmental science in general and climate-related science in particular, and datasets have often been the first target for attack in the guise of "re-interpretation." Originators are thus understandably wary of putting their data on full and open display.

Allchin's view is that this threat will best be met by adopting standardized processes and protocols covering every stage of the data-management chain, from observation and download, through transmission, validation and storage, in much the same way that crime-scene evidence is handled by forensic investigators. Noting work by a committee of the US National Academy of Sciences, Allchin offers three "pillars of data husbandry" as sound principles of data management:

- integrity: the truth and accuracy of data, driven in part by adherence to rigorous standards in observation and validation
- access: the degree to which datasets are made openly available
- stewardship: the need for datasets to be recognized as major assets and therefore that provision be made for their long-term preservation.

Such a degree of transparency, Allchin argues, will help to enhance the credibility of all data, particularly in collaborative research networks, and thereby provide a firm foundation on which to build new hypotheses and models.

*"Amateurs talk theories; professionals talk data."*
—MICHAEL ALLCHIN, DATA MANAGER,
IP3 & WC$^2$N RESEARCH NETWORKS

*John Pomeroy*

With such a broad range of data as that generated by the two research networks, seeking a "one size fits all" solution for archiving was not a realistic option. In IP3 alone, for example, there are some 30 million individual values, from 425 separate instrumentation arrays, at 45 stations in the seven main research basins. Projects which aim to develop systems able to accommodate large volumes of disparate data do exist, but the resources required come close to the entire budget for either network. The challenge, therefore, was to build these archives in a manner appropriate to the data being published and the technical resources available, yet also enable indefinite availability without ongoing support once the networks had shut down. The answer was to adopt relatively "low-tech" methods for storing datasets and metadata, supporting downloadability of these from the network websites and thereby making them available to the broadest possible audience for as long as those sites exist.

The same datasets, however, have also been made available

to the Water & Environmental Hub (the WE-Hub), an initiative of CYBERA (a Calgary-based organization which seeks to advance cyber-infrastructure in Alberta). This $2-million project is aiming to develop an advanced, cloud-based platform to support the archiving and visualization of a wide range of data, with the intention of evolving from a research and development project to a fully operational platform.

The hope is that future generations of researchers will know where the datasets collected during these two research initiatives are located and will trust in their veracity, and that the data should therefore continue to contribute to future understanding of the hydrology and climate in Canada's colder regions well past the closure of both networks.

This is a point not lost on Dr. Steve Liang, leader of the Geospatial Cyber-infrastructure for Environmental Sensing, or GeoCENS. This project, also funded by CYBERA as well as by CANARIE (Canada's Advanced Research & Innovation Network), is a collaborating partner with IP3.

Hydrologic, atmospheric and other geographic data are gathered with sensors. Liang points out that remote sensors of many kinds are located everywhere in our environment, that many such sensors are already equipped with communications, and that the cost of embedding sensors and communications devices into other technology is decreasing rapidly. However, the vast majority of sensor arrays currently operate in isolation from all others, working only for the purposes of their respective owners. Liang makes the point that the value of any given digital network, as exemplified by the social-networking websites of our age, is proportional to the square of the number of links in the system. By establishing links between sensor subnetworks, it should in theory be possible to extend the principle of "crowd-sourcing" to the observation of natural phenomena and so support a dramatic increase in the availability of environmental data.

Liang's plan for GeoCENS is that it should provide a comprehensive and integrated host for environmental sensors in one global network, supporting broad collaboration on information related to the physical state of the earth's surface and atmosphere. In the same way that contributors continually update Wikipedia entries or maps on the Open StreetMap website, GeoCENS invites data-gatherers around the world to contribute and share observations of natural phenomena of all kinds, whether now or at any stage in the past and at any point on the earth's surface. If Liang's ideas of sensor proliferation and their connection to GeoCENS were to play out, the density, resolution and frequency of observations available to scientists would increase exponentially. This would greatly assist in the parameterization of the broadest range of hydrological and atmospheric conditions, in turn supporting more accurate and reliable prediction.

The GeoCENS interface is similar to the familiar Google Earth and other web-based mapping utilities. But unlike those sites, zooming in to finer resolution of the major geographical features at GeoCENS reveals not businesses, photographs and points of interest, but pinpoints indicating the locations associated with stored datasets. These might represent any sensor of natural phenomena, from an automatic weather station to, potentially, a daily update from an amateur meteorologist's home weather station or perhaps a backcountry skier's snow-profile pit. Users may interact with the system along the lines of established social networking websites, commenting on the data, station suitability and other details.

The GeoCENS mission is thus to enable scientists to collaborate, and to access and share sensor data, in a way and at a scale never previously possible. Liang's hope is that the system could provide a major input stream to support open collaboration, allowing data providers to create new sensors and publish associated datasets in immediately searchable and accessible formats, thereby enabling nearly real-time use. The system could also enable users to control sensors and to be notified when specific sensing tasks are completed. The GeoCENS sensor web could thus, in principle, offer the possibility of dynamic reprogramming of individual sensors or subwebs within the network for specific functions, perhaps to respond to regional environmental or climatic crises during extreme flood or weather events. Liang contends that as the quality and range of capacity of the sensors improve, they will evolve into intelligent systems which will offer immensely greater flexibility than today's "dumb" data-collectors.

Though Liang didn't say it this way, the development of these kinds of capabilities will turn monitoring from a passive into an active scientific activity. What he did say, however, was that GeoCENS offers a step toward the goal of creating ever-broader collaborative global data sharing and research networks.

While members of the IP3 and WC²N networks see potential in the GeoCENS concept, they also recognize the challenges Liang and his colleagues face in creating a system of this magnitude that actually provides accurate and useful information. The project faces the same problem that currently haunts all open-source digital networks on the worldwide web, namely that information or data is only useful if it is known to be reliable and trustworthy in its provenance, and in a crowd-sourced system this quality measure is largely unverified. An analogy on Wikipedia is the reportedly high frequency at which pages relating to politicians are altered, by supporters and detractors alike. If there is no constraint on who is able to make edits, how does one assess the validity of the information at any particular time? What Liang is offering is a dream that will only make progress toward fulfillment if scientific users buy in to make it work. The contribution of these networks in this context is to be found in the provision of the IP3 dataset to the WE-Hub, for which the GeoCENS architecture is providing an important component.

Back at the conference in Red Deer, it seemed impossible to everyone in the room that work of this importance was about to be halted because the government of Canada saw no value in continuing to fund climate-change-related hydrological, cryospheric and atmospheric research. The IP3 and WC$^2$N networks were on the verge of critical breakthroughs, not just in improving parameters that will make prediction more reliable but in building systems that will dramatically allow us to better understand earth processes in real time. It was heartbreaking to realize that this critically relevant work would come to an end in only a few months time simply because the current political party in power in Ottawa did not believe that climate change was a serious enough matter to warrant further research.

## HYDROLOGY, GLACIOLOGY AND SNOWMELT

On the second day of the Red Deer workshop, Dr. Pomeroy introduced some of the research outcomes that have been derived through the use of the cold regions hydrological model. He began with research he and his University of Saskatchewan colleagues had undertaken in association with the U of C's Biogeoscience Institute at Marmot Creek in the Kananaskis region in the Rocky Mountains of Alberta. Those findings were that if the canopy of the mountain forest remains intact, there is virtually no increase in snowmelt, even with a 4°C mean annual temperature increase. Remove the forest canopy, however, and snowpack drops by half and spring snowmelt advances by a month. From these observations it can be seen that forests function to self-regulate local hydrology and do in fact slow and moderate the effects of climate change as many suspected.

Pomeroy then talked about the related work of Masaki Hayashi of the University of Calgary, the leader of a six-member IP3 team studying hydrological storage and groundwater pathways in alpine headwaters in the watershed of Lake O'Hara in adjacent Yoho National Park in British Columbia. This research group wanted to find out where and how much groundwater was stored in the region beneath Opabin Glacier, called the Opabin sub-basin. They also wanted to know how long groundwater was stored in this sub-basin and how all of this information could be represented in basin hydrology models.

Between April 16 and 21, 2007, Hayashi and his research team conducted hundreds of snow depth and density measurements in the sub-basin in order to calculate the snow-water equivalent of the snowpack. They also employed a laser range finder – which they called their "poor man's LiDAR" – to model the depth distribution and calculate the snow equivalent on surrounding slopes too steep to measure directly with conventional probes. They utilized what is known as the ArcGIS radiation tool to estimate the sun's impact on the snowpack and electrical resistivity to estimate groundwater flows through dry and wet moraines, debris-covered ice, degrading permafrost, and bedrock. In addition – and with Parks Canada's permission – they also released 44 kilograms of salt into a small pond of accumulated meltwater to determine how groundwater travelled under the surrounding talus slopes. Utilizing a groundwater flow model, Hayashi's team then simulated how the water stored in a tarn, or small pond on the moraine, dispersed through groundwater channels.

The researchers determined that groundwater storage time in the talus was in the order of less than a week. They also discovered that loose sediments were also carried in high concentrations through groundwater systems. All of the parameterizations established through this painstaking research are now incorporated in a new, coupled surface and groundwater model that simulates basin outflow. This had never been done before in the headwaters region of the Canadian Rocky Mountains protected as national parks.

Pomeroy also talked about the work of Dr. Scott Munro of the University of Toronto, who is famous in the glacier research community for decades of work on glacier mass balance on Peyto Glacier in Banff National Park. Like John Pomeroy, Dr. Munro also explored processes, parameterization and prediction in tandem to derive meaning and value from a new 2002 to 2007 hydrometeorological dataset. His work on parameterization focused on advanced characterization of the interactions that take place between the weathering crust of the melting snowpack, the cooling layer of air above it, and the warmer air layers above the cooling layer. Munro hypothesized that due to the heat flow barrier of the glacier and windspeed maximum, air adjacent to the surface of the ice on Peyto Glacier cools as it accelerates down-glacier. In accordance with this supposition, Munro further hypothesized that the action of wind eddies at maximum wind speed height (considered as a single vector value) did not affect air temperature. His third hypothesis was that, due to the ongoing weathering of the crust, there is significant delay in meltwater flow response at the glacier's surface during the course of any given day.

What Munro was describing was warming and modest increase in meltwater streamflow due to sudden warming associated with intermittent boundary layer collapse in the atmosphere near the glacier surface. At an IP3 conference in Whitehorse, Munro had called these boundary layer collapses "hot flashes," which certainly got the audience's attention if only because they could immediately imagine how the media might respond to the notion of "menopausal" heat pulses emanating from the disrupted surfaces of aging, soon-to-be-disappearing glaciers.

Munro has gone on to parameterize these processes and to model basin discharge predictions based on what he has learned. In terms of his final hypothesis, Munro has determined that based on 25-metre data grids available from satellite and remote sensing from aircraft, the current modelling approach to parameterization and prediction may be the best that researchers can do for now. It is not perfect, he said, but certainly a start.

Munro's analysis of how melt on glaciers can be parameterized, Pomeroy explained, dovetails with the work of Danny Marks of the Northwest Water Research Centre of the US Department of Agriculture's Agricultural Research Service in Boise, Idaho. Dr. Marks and his five-member team are conducting research on hydrologic modelling over snow-dominated mountain basins. They are attempting to advance the accuracy of parameterization by coupling an already well-developed model called ISNOBAL with a physics-based model developed at Pennsylvania State University.

## DANNY MARKS

Danny Marks is a hydrologist who specializes in snow and cold-season hydrology of mountain regions. Marks and his soil scientist colleague Mark Seyfried are at the United States Department of Agriculture's Agricultural Research Service Northwest Watershed Research Center in Boise, Idaho, which manages the Reynolds Creek Experimental Watershed in Idaho's Owyhee Mountains, near Boise. The Owyhee Mountains are part of the Great Basin of the intermountain western United States. Reynolds Creek is the USDA's premier mountain observatory.

Over the past five decades, mean temperatures rose 2 to 3°C, according to Agricultural Research Service data from the Reynolds Creek watershed. Warming has been found across a range of elevations, with the greatest increases occurring in minimum daily and seasonal temperatures. The impact of this has been less snow and more rain, again at all elevations, but with the most profound changes occurring at lower altitudes, which saw a shift from 44 per cent snow to 20 per cent snow.

Fifty years of records in the western mountains of the United States show that while it is now significantly warmer, total annual precipitation in the basin has not changed. But a system that was once dominated by winter snowfall now experiences a mix of rain and snow, with more streamflow in winter and less in spring. As a result, there is less water for ecosystems and agriculture during the spring and summer growing season. These changes make forecasting and managing western water

resources more difficult and present a serious challenge to agriculture in the region.

## THE ISNOBAL MOUNTAIN SNOWMELT MODEL

ISNOBAL owes much of its development to Danny Marks and a colleague named David Garen, who works with the USDA in Portland, Oregon, where the model was used to simulate streamflow forecasting in the Klamath River basin. The model also acquired some of its sophistication through work on that basin undertaken by Tim Link of Oregon State University. John Pomeroy has also collaborated in the model's development through research he and Marks did together in the mountains of Colorado.

The ISNOBAL model (see page 131) is designed to simulate both the development and melting of seasonal snow cover in mountain basins. To do this, modellers require running measurements of air temperature, humidity, wind speed, precipitation and solar and thermal radiation estimated at intervals of three hours. Unfortunately, the full range of parameters that affect climate conditions at various altitudes and aspects of exposure that affect the deposition and melt of snow in mountain basins is seldom monitored. Danny Marks and the rest of the IP3 research group aim to develop improved tools and methods that can use limited data and relatively simple methods to produce computer-simulated images of these processes that are reasonable approximations of actual conditions.

Like other models, ISNOBAL needs to take into consideration a great many parameters in order to be able to simulate the complex processes that occur every day in nature. ISNOBAL is a grid-based distributed energy balance snowmelt model that encompasses four different parameter sets. There is a set of five "state" variables, which include snow depth, snow density, snow surface layer temperature, average snow cover temperature and average liquid water content. There is also a set of three mass flux parameters, including evaporation, melt and surface water input. And there is a set of seven

## THE INGREDIENTS OF A SNOWMELT MODEL

Like other mathematical models, ISNOBAL needs to take into consideration a great many parameters in order to simulate the complex processes that occur every day in nature.

All of these heat flux values influence rates of sublimation, evaporation and condensation in or on the snowpack. For discontinuous snow cover in open environments, local advection, or the transfer of heat as a result of the movement of energy from bare ground to snow patches, is also an important parameter in determining the energy it takes to create and melt the snowpack.

The table below illustrates what you need to know to model snow processes.

| Class of Variable | Defining Parameters | Sub-Parameters |
|---|---|---|
| state variables: | snow depth | |
| | snow density | |
| | snow surface layer temperature | |
| | average snow cover temperature | |
| | average liquid water content | |
| mass flux variables: | evaporation | |
| | meltwater input | |
| | surface water input | |

| Class of Variable | Defining Parameters | Sub-Parameters |
|---|---|---|
| forcing variables: | incoming thermal radiation | |
| | net solar radiation | |
| | air temperature | |
| | vapour pressure | |
| | soil temperature | |
| | wind speed | |
| | precipitation | |
| radiation and energy variables: | exothermic radiation | visible radiation |
| | | near-infrared radiation |
| | clear-sky radiation, corrected for topography and cloud cover | |
| | global and diffuse radiation | direct radiation |
| | | sub-forest-canopy direct radiation |
| | | net snow cover radiation |
| | | net diffuse radiation |
| | | sub-canopy diffuse radiation |
| | energy parameters | ground heat flux |
| | | advective heat flux |
| | | latent heat flux |
| | | sensible heat flux |

forcing variables, including incoming thermal radiation, net solar radiation, air temperature, vapour pressure, soil temperature, wind speed, and precipitation.

A fourth set of variables has to do with radiation and heat energy. No one ever said nature is simple.

The radiation parameters are divided into exothermic radiation, which includes both visible and near-infrared radiation; and clear-sky radiation, which has to be corrected to account for topography and variance in cloud cover. Measured global and diffuse radiation is subdivided into direct radiation, sub-forest-canopy direct radiation, net snow cover radiation, net diffuse radiation and sub-canopy diffuse radiation.

If all that hasn't made you dizzy yet, there is a final subset of energy-related parameters, which, like all the rest of these variables, has to be measured in order to ensure consistency between modelled results and actual observations in nature. This group has to do with heat transfer, and each of them in its own way influences rates of sublimation, evaporation and condensation in or on the snowpack.

The first of these, ground heat flux, is the amount of heat flowing from the earth's interior through a given unit area at the surface. This is a particularly important parameter in cold regions.

Advective heat flux is the transfer of heat energy from bare ground to snow patches. For discontinuous snow cover in open environments, local advection is an important measure of the energy it takes to create and melt the snowpack.

Latent heat flux is the heat exchanged due to phase changes such as those that occur through evaporation, sublimation or refreezing.

And finally, sensible heat flux is the flow of heat resulting from convection, or the transfer of heat from one area to another as a result of thermally induced differences in the density of the snowpack caused by warming from the bottom and cooling from above or vice versa.

All of these parameters are known to change as the climate warms.

In order to arrive at the most accurate simulations, models are usually nested one inside the other so that each process is accurately simulated within the larger framework. Rain, snow and radiation inputs are filtered serially through an evapo-transpiration loss model, a canopy interception model, a snowmelt model, a one-dimensional unsaturated zone model for each grid element, and finally a saturated zone model for each of the same grid squares. Each of these separate model elements operates under its own algorithms.

In order to make each additional run of model results more and more faithful to what actually happens in nature, IP3 researchers are constantly adding new parameters to the process. Marks and his colleagues linked the Penn State Integrated Hydrology Model, known as PIHM, to the existing suite of ISNOBAL submodels with the hope of learning more about how below-ground processes interact with snow processes to control the hydrology of snow-dominated mountain basins.

Their conclusion was that ISNOBAL could be made to yield more faithful results by incorporating the Penn State submodel, provided that better groundwater data were available that would allow for multi-year simulation of groundwater storage. Just as Scott Munro had reported about the advancements he had made in parameterization on Peyto Glacier, Marks has pointed out that the results are not perfect. They never are, but the advancements were perhaps the first links between a detailed snow hydrology and a groundwater model and were a major step in a needed direction for understanding and predicting mountain water resources.

## MODELLING THE NORTH AND THE PRAIRIES ...

From the Rockies, Dr. Pomeroy moved his discussion of the cold regions hydrological model northward to discuss its application in boreal and arctic situations. The challenge in the North is how to model in basins where there is little or no monitoring. This requires the transferring of parameters from elsewhere to places where there are no gauges. Pomeroy demonstrated that with careful parameterization and bottom-up and top-down modelling, it is possible to represent and conceptualize landscape and hydrological variability in subarctic environments.

The subject then moved to the prairies and important work undertaken by Pomeroy and his team on Smith Creek in Saskatchewan. Here, localized hydrology is affected by poor drainage. Many areas possess storage in landscape depressions but contribute nothing to streamflow. Smith Creek is underlain by heavy glacial till. Because of the variable conditions that affect the streamflow contributing areas, the presence of wetlands, frozen soils and evapotranspiration from established plant communities, Smith Creek is a classic research basin. The model demonstrated that different kinds of land use have very different effects on drainage. Forest conversion reduces drainage by 10 per cent. Agricultural conversion increases drainage by 20 per cent. Wetland drainage increases streamflow by 70 per cent. Wetland restoration, however, was seen to reduce streamflow by only 30 per cent.

The implications of this research are significant. Despite a huge commitment, wetland restoration by organizations such as Ducks Unlimited and others is not keeping up with agricultural conversion and wetland drainage. Pomeroy and his colleagues discovered that streamflow volumes are far more sensitive to wetland drainage and restoration than they are to changes in forest or agricultural land use. In the prairie basins in the Canadian West, wetlands play a far more important role than imagined in the past. They may in fact hold these ecosystems together.

With each of these discussions, however, we arrived back at the same challenge. We need high-resolution information on catchment characteristics to accurately model prairie snowpack, soil moisture and streamflow. What this means is that if you don't have accurate monitoring – or worse yet, if you don't have any monitoring at all – then it becomes even more imperative that the surrogate information you collect from elsewhere be as accurate as possible relative to where it was collected. If it isn't, then the margins of error in the first set of calculations will be amplified when they are aggregated in the second set.

## ... AND CONFRONTING THE POLITICS OF PARAMETERS

At this point, a number of participants could no longer contain themselves. One pointed out that the list of parameters the watershed alliance where he lived used to compile their 400-page "state of the watershed" report did not respond to or in any way take into account many of the parameters and considerations Dr. Pomeroy had indicated as essential to the effective operation of the cold regions hydrological model. An animated discussion followed during which it was revealed that "state of the basin" reports were often defined by parameters acceptable to those with vested interests in that specific basin, not by objective scientific criteria. It was pointed out, for example, that the agricultural sector had lobbied successfully to have the influence of intensive livestock operations removed from a number of state of the basin reports, and that climate change effects have been ignored in others. The vigorous discussion that followed made it clear that science still has a long way to go to establish its relevance in the ongoing development of water policy.

## COMING TO GRIPS WITH EXTREME COMPLEXITY

John Pomeroy's course on observation and simulation through modelling made it obvious that natural systems are truly telekinetic in that even at a distance a slight change in the value of any one parameter can immediately affect many others. Each level of complexity leads to another, with the total system becoming complicated almost beyond comprehension. It also became apparent that almost anything around us could be considered a hydrological response unit, which left some of the participants with the sense that we are blundering around our landscapes without any real knowledge of the impact we're having. It became obvious as well that we have little baseline information upon which to measure change.

The hydrology of the Canadian West is complicated enough to make one's brain explode. John Pomeroy challenged the participants in his course to think about embedding models of this complexity within a much larger climate context where the number of parameters and variables expand by two orders of magnitude. Issues related to water quantity and quality are embedded within the hydrological cycle. If we can understand one, then maybe we will be able to understand the other.

By understanding the characteristics of the parameters that define nature in a given place, it becomes possible to follow water to the centre of natural ecosystem function. To understand the parameters by which water functions in the world is to know the relationships that define natural design. Recognition of the fundamental parameters that circumscribe the means by which water circulates through life and through the world is to glimpse the heart of nature itself. Parameters can be organized into algorithms, or sets of relationships, that allow us to see how natural processes actually function individually and in relation to one another. These algorithms can be used in serial combination to create credible, realistic models of how the hydrological cycle permeates our reality.

The resulting mathematical models can then be used to improve precipitation forecasts for western Canadian farmers, to better predict the risks of flood and drought, to determine how much surface water and groundwater will be available to supply our growing cities, to make sure there is enough water for industry and ensure that enough stays in our rivers and streams to support vital aquatic ecosystem health, and to sustain our aesthetic and recreational needs. Together all of these needs amount to the greatest prize of all. The potential value to society of defining the parameters that make water do what it does in a given place, and of linking established parameters to larger natural processes such as climate, cannot be underestimated. What the researchers participating in the Improved Processes Parameterization & Prediction Network (IP3) and the Western Canadian Cryospheric Network (WC$^2$N) are striving to discover is nothing less than the fundamental principles by which water facilitates larger earth system function, from which they hope to derive the ultimate algorithm, the $E=mc^2$ of true sustainability.

# CHAPTER 2

## Ice Matters:
## The State & Fate of Canada's Glaciers

The research outcomes generated by the IP3 and WC²N research teams suggest there is a great deal an interested person can learn from being open to emerging understanding of climate science. Anyone who has struggled successfully to comprehend the mechanics of climate change as it relates to global atmospheric dynamics will appreciate even more the complexity of our earthly life-support systems.

To gradually be able to visualize the fragile, highly interactive and dynamically responsive glory of the atmosphere on one's own terms can be a life-altering experience – especially for someone interested in sharing the world with others. The first true glimpse of just how sensitive our planet's atmosphere is to orbital eccentricities, material inputs, minor shocks and human influences can take your breath away. It is one thing to learn about this from others, quite another to discover it for yourself.

It is stunning to realize how quickly and completely the properties of our atmosphere can change just by adding heat or small amounts of various gases and aerosols. At the moment, anthropogenic emissions are causing the equivalent of 1.5 to 2.0 watts of warming per square metre of the earth's surface, which is about the warming effect you would expect to get on the ground beneath two outdoor Christmas lights. What is astounding is that the earth's system is so sensitive and so utterly and completely interconnected that even that small amount of extra heat will eventually warm our oceans.

The mind slows at the realization of how much an entire global system can be altered by the addition or subtraction of even a few parts per million of substances that already compose it. The moment one can see in their mind's eye how such additions and subtractions can change the seasons or bring on or delay ice ages, the world is no longer the same.

Suddenly what you are seeing when you look upward into the blue of the sky is not just air but the suspended residue of every geological event and the cumulative exhalation of every ecological process that has ever taken place on or near the surface of the earth and its oceans since the beginning of the world.

One is at once dazzled and humbled by the realization that our mountain ecosystems contribute to and are in turn made possible by interpenetrating, self-regulating global processes which are ultimately represented in the composition and resulting conditions created by our planet's thin and fragile atmosphere. Life's existence in the high Rockies is suddenly not only more amazing than we have realized, but likely more amazing than we can presently imagine.

So how does one get their head around something this complicated? One way is to pick a single strand and follow it into the larger weave of climate circumstance. Because liquid water and snow and ice respond directly, visibly and measurably to temperature, water is an obvious choice as a strand we can easily follow. If we follow what is happening to our water, it will tell us what is happening to our climate. We can begin following climate's effect on water by observing what climate does to ice. That is exactly what researchers in the Western Canadian Cryospheric Network have done.

## WC²N

Brian Menounos was the coordinator and a lead investigator of Western Canadian Cryospheric Network, a research project that involved scientists at six Canadian and two US universities between 2006 and 2010. This network came to be known in the scientific community as WC²N. The objective of this research was to clarify the fate of glaciers in Western Canada.

In the courses he teaches at the University of Northern British Columbia in Prince George, Dr. Menounos points out that glaciers are natural climate stations. Under the aegis of the Canadian Foundation for Climate and Atmospheric Sciences, Menounos and his WC²N team conducted research that indicated the vulnerability of downstream ecosystems to changes in the balance between winter snow accumulation and summer melt of western Canada's remaining glaciers. This work underscored the fact that rising electricity demand in Western Canada is on a collision course with decreasing stream discharge in western rivers.

As a foundation for this work, researchers in the WC²N collaboration documented climate variability and changes in glacier extent in the North Pacific over the last 400 years. They also detailed meteorological processes and the links of those to the nourishment of glaciers as determined by mass balance, which is defined as the net annual gain or loss of the mass of a glacier when snowfall and refrozen rain are balanced against the wasting effects of melting and sublimation. Finally, researchers used the information generated by analysis of the loss of glacier extent over the past four centuries and patterns of contemporary nourishment of glaciers to project how glaciers will respond to the climate that is projected to exist over the next 50 to 150 years. This project, which engaged ten undergraduate assistants, 18 master's students, nine doctoral students and seven post-doctoral fellows, used a combination of techniques to determine the state, and then project the fate, of the remaining glaciers in the mountain West.

Historical maps were compared with contemporary oblique and aerial photography and satellite imagery over a vast region of the mountainous Pacific Northwest, including all of British Columbia and east to the front ranges of the Rocky Mountains in Alberta. From the Landsat series of satellite maps, $WC^2N$ researchers established the extent of glaciers over an area many times the size of Switzerland.

Through patient analysis and careful consideration of all the analytical tools available to them, $WC^2N$'s Brian Menounos and Roger Wheate and their colleagues have been able to understand how climate affects glacial ice in the mountain West. In the face of limited support for hydrological and climate research, this team has been able to figure out the best way to conduct scientific research on glaciers without having to rely on costly direct monitoring of what is actually happening to our country's ice. The principal tool they used to do this is a simple one. They compared what exists today as evidenced by satellite and aerial photographs to what earlier maps depicted in terms of glacial extent. Through these comparisons they were able to roughly determine how much the area covered by glaciers in the mountain West has changed over time.

Menounos and Wheate, along with Matt Beedle and Tobias Bolch, compared glacial extents in the western Cordilleran region between 1985 and 2005 as revealed by Landsat analysis of glacial area. They divided the Western Cordillera into nine regions: the St. Elias Mountains, the North Coast Mountains, the Central Coast Mountains, the Southern Coast Mountains, the Northern Interior, the Southern Interior, the Northern

Rocky Mountains, the Central Rocky Mountains and the Southern Rocky Mountains. As one would expect given the vast differences in precipitation and latitude of the peaks in each of these nine regions, the rate of change in each differed dramatically (see map on page 131).

With these analyses as a foundation, researchers then compared the number and area of the glaciers that existed in 2005 with similar satellite data and photographs taken from aircraft from 20 years before and with information derived from topographical maps created in the 1920s. Out of this research came some astounding discoveries.

## WC²N AND THE CANADIAN ROCKY MOUNTAIN PARKS WORLD HERITAGE SITE

The frozen water we may wish to follow first in our quest to understand climate's effect on western water is that which exists as icefields and glaciers in the Canadian Rocky Mountain Parks World Heritage Site. Many readers will have visited this region of Western Canada and will be able to imagine the landscapes and the glaciers we will describe. This renowned 24,000-square-kilometre protected area straddles the Continental Divide, which marks the boundary between Alberta and British Columbia. Recognized by UNESCO as one of the most spectacular examples of temperate mountain landscapes on earth, this World Heritage Site encompasses Banff, Jasper, Kootenay and Yoho national parks and Mount Robson, Hamber and Mount Assiniboine provincial parks, which taken together are widely held to form one of the most strikingly beautiful regions on the planet.

Through the work of the Western Canadian Cryospheric Network, Canadians are rediscovering the fact that, in addition to being one of the most significant and enduring public policy achievements in the history of the Canadian West, these protected upland areas are a vital source of increasingly scarce western water.

Parks Canada worked closely with WC²N researchers to learn as much as possible about what the glaciers, rivers and headwaters streams in the Rocky Mountains are telling us about how changes to our climate might affect western water supply. What they learned was that national parks and other protected places appear to possess a seemingly unlimited capacity to provide new insights and discoveries to each subsequent generation that experiences them. In the same way that pristine natural landscapes provide a baseline for measuring and evaluating changes to our mountain ecosystems, upland protected areas also provide a foundation for measuring changes in the hydrology of the Canadian West. As a result of the WC²N findings, Parks Canada managers learned that it is not just water itself we are protecting in our mountain parks. We are protecting the places and to some extent the processes that gather water from the sky and make it available when and where we need it in the Canadian West. In other words, we are protecting the hydrological cycle.

We now know that the scale and ecological stability of this protected area actually slows and moderates climate-change impacts, not only regionally but in western North America as a whole. That we have protected such an important area of the mountain West – as we shall see – may also be one of our most important means of adaptation to accelerated warming that our receding glaciers warn us to expect. In a very real sense, what we have saved may very well save us.

## GLACIER RECESSION: THE GLOBAL AS REPRESENTED IN THE LOCAL

According to the World Glacier Monitoring Service, there are about 160,000 glaciers on this planet, covering in total an estimated 685,000 square kilometres. In most places in the world where glaciers have been studied, however, efforts to monitor their health have not been continuous, resulting in lengthy gaps in the data collected. Globally there are only 39 glaciers that have been studied for more than 30 years, and only 30 "reference" glaciers that have been subject to continuous measurement since 1976. One of these, Peyto Glacier, is in the Canadian Rocky Mountain Parks World Heritage Site (see research station photo on page 133). It has been the object of research conducted by the Geological Survey of Canada for more than 40 years. This research continues today under Dr. Michael Demuth, who heads the glaciology division of the Geological Survey of Canada in Ottawa (see photo on page 132).

Research conducted on the world's 30 "reference glaciers" indicates a cumulative global loss in glacial mass of 20 per cent in the 60 years between 1945 and 2005, which suggests that a great deal of the world's ice has already become water. All studies indicate that dangerously warm years such as we experienced in the northern hemisphere in 2003 are likely to become more common. We know what the extreme temperatures of that year did to the Alps, which in places lost 10 per cent of their glacial mass in only one summer. We are only now beginning to understand the impacts here.

In 2003 extreme increases in minimum temperature as reported from recording stations in the Rockies were in the range of 7 to 8°C above the established mean summer daytime and nighttime temperatures in this part of the mountain West. Mountain guides crossing the major western Canadian icefields reported that for the first time in living memory snow on the uppermost accumulation zones of these icefields was melting even in the middle of the night. The fear, of course, is that extreme temperature events of this magnitude could push our climate system out of equilibrium. The problem, as climate scientists have already indicated, is that we are not ready for these kinds of changes (see photos on page 134).

Between 1975 and 1998, glacier cover in Alberta's North Saskatchewan River basin, as measured by area, decreased by approximately 22 per cent. During the same period, glacier cover as measured by area in the South Saskatchewan basin decreased by 36 per cent (see graph on page 136).

Some three million people live in these two river basins. The ice volume wastage estimated for the North Saskatchewan basin expressed as an annual average is equivalent to the amount of water used by approximately 1.5 million people. When there is no ice left to become water, the growing needs of water-reliant populations, industries and agricultural sectors will exert ever-increasing pressure on water availability.

No one who knows the mountain West disputes the fact that our glaciers are disappearing. Witnesses have also observed that the loss of glacier ice is not limited merely to the length of glaciers but more significantly to their depth. Our glaciers are down-wasting at a very rapid rate, but until the work of IP3 and WC²N, only limited information existed about how quickly these changes were occurring and what they might mean for water supply or to ecosystem function.

## GLACIER LOSS IN THE MOUNTAIN PARKS

WC²N credits post-doctoral fellow Tobias Bolch for the careful analyses that yielded the information that made the glacier inventory possible. As a result of this work, WC²N researchers estimated that on average the annual rate of area loss in the mountains of British Columbia was about one-half a per cent (0.54 per cent ±0.15 per cent). The average rate of annual area retreat of glaciers in Alberta was 2.5 times as great, or about 1.25 per cent a year (1.27 per cent ±0.17 per cent). Over the entire cordillera, including the St. Elias Range, the average annual rate of area retreat was 0.55 per cent, which reflects the fact that there are many more glaciers in British Columbia than in Alberta. The shocking news, however, is that while there may be more glaciers in British Columbia than in Alberta, glaciers are rapidly becoming smaller in total area in both provinces. Even more alarming is the fact that glaciers are not just receding but vanishing altogether.

According to WC²N glaciologists, Jasper National Park may have had 554 glaciers in 1985. Twenty years later, as many as 135 of those had disappeared. If these estimates are correct, the area of the park covered by glacial ice had diminished by 13 per cent.

In 1985, the year Banff National Park celebrated its centennial, it appeared that Banff had 365 glaciers within its boundaries, covering 625 square kilometres. Evidence suggests that 20 years later 29 of those glaciers had disappeared and the area of Banff National Park visibly covered by ice had been reduced by some 19 per cent to just over 500 square kilometres.

What happened between 1985 and 2005 in Banff and Jasper, however, was different from what happened in Glacier, Yoho, Kootenay and Mount Revelstoke national parks. The number of glaciers in each of the latter four parks actually grew between 1985 and 2000. The reason for this was not that climatic conditions were bringing glaciers into existence, but actually quite the reverse: glaciers were diminishing in area so rapidly that individual ones were breaking in two or even three smaller, isolated masses. While some who wish to deny climate change will take the increased number of glaciers in some areas as evidence that warming is not occurring, it should be noted that although the number of individual glaciers increased, the total area they covered continued to diminish. It is interesting to observe that the median area of glaciers in Canada's western mountain national parks is less than two square kilometres. Small glaciers disappear more quickly and that is what happened between 2000 and 2005 (see images on pages 134 and 135).

## GLACIER LOSS IN ALBERTA'S PROVINCIAL PARKS

A similar decline in glacial cover has been observed in the mountain regions outside of the western national parks, especially on the eastern slopes of the Rocky Mountains. Of the 41 glaciers that visibly existed in the Willmore Wilderness in 1985, 14 may have disappeared, reducing the area covered by glacial ice by 19 per cent. There were 28 glaciers in the White Goat Wilderness in 1985. Now there appear to be only 20, and the area visibly covered by ice has been reduced by some 30 per cent.

Because the 24 glaciers that remained in the Siffleur Wilderness in 1985 were found only in high, north-facing cirques and on average still cover only about half a square kilometre in area, only one of these glaciers has disappeared in the past 20 years.

In the Kananaskis, however, it is the remaining lower-elevation glaciers that have been the hardest hit by warming. Of the 25 that existed there in 1985, only 21 remain. The area covered by glacier ice in the Kananaskis appears to have diminished by 35 per cent in 20 years. In the Spray valley there is only one glacier, the Robertson, and though it has not disappeared, it lost 40 per cent of its area between 1985 and 2005.

The combined area of glacier-covered landscape in the Willmore, Siffleur, White Goat, Kananaskis and Spray areas appears to have been reduced from 154 square kilometres in 1985 to 122 in 2005. This means that a total of 32 square kilometres of ice cover may have been lost in these areas in only 20 years.

In the western mountain national parks alone, the total

glacier-covered area appears to have been reduced from 1870 square kilometres in 1985 to 1560 in 2005. This means that the character of up to 310 square kilometres of mountain landscape completely changed in only 20 years (see, for example, the map on page 137).

Working under the supervision of Brian Menounos and Roger Wheate, doctoral student Matt Beedle also undertook research at Castle Creek Glacier in the Cariboo Mountains of northern British Columbia. Here they are using push moraines at the snout of the glacier as a climate proxy. In work done in collaboration with Brian Luckman, Beedle and his team determined that Castle Creek Glacier had receded 701 metres between 1959 and 2007. They observed that average glacial recession was 14.3 metres a year in the mid- to late 1970s. Recession in the 1990s, however, accelerated to over 40 metres a year, confirming again the rapid rise in summer temperatures in the northern hemisphere in the final decade of the 20th century (see photo on page 138).

## EXTENDING THE RECORD FURTHER BACK IN TIME

Good science employs the broadest range of analyses to confirm the validity of findings. Working with Brian Menounos and Roger Wheate, Christina Tennant used the opportunity afforded by $WC^2N$ funding to further explore the huge, largely unexploited potential of using historical maps as a foundation for examination of glacial recession in the mountain West. In a presentation called "Eight Decades of Glacier Change in the Canadian Rockies," given at an IP3–$WC^2N$ conference at Lake Louise in October of 2009, Tennant demonstrated how photo-topographical work done during the surveying of the Alberta–British Columbia boundary between 1913 and 1924 could be reinterpreted to yield information about the changing state of glaciers between 1920 and 2005.

After explaining the history of the Boundary Survey, Tennant enthusiastically demonstrated how maps from the 1920s and later surveys could be "geocoded" so as to allow them to be compared with contemporary aerial photographs and satellite imagery. The comparisons could be used to derive information about how quickly remote, unmonitored glaciers along the Great Divide of the Rocky Mountains have been receding over the past eight decades.

From this analysis, Tennant and her colleagues made some stunning discoveries. During the period between 1920 and 1985 the minimum elevation at which glaciers were found in this region of the mountain West rose by 150 metres. Between 1985 and 2005 the minimum elevation at which glaciers were found rose by an additional 50 metres in only 20 years. During the same time, the snow accumulation season shortened, and not only has the ablation, or melt, season continued to grow longer, but melting is occurring at higher altitudes during the ablation season. It is no surprise that small glaciers experienced the greatest percentage of area loss. (See, for example, the photos on page 141.)

Tennant concluded by emphasizing the potential value of Boundary Survey maps in glacial research. Though she observed that some glaciers were imprecisely drawn on some of the maps, on the whole they were surprisingly useful in efforts to compare how much water was banked in ice in the past compared to now. Tennant also indicated that her work had just begun. It was her intention to try to get more information about the volume of ice we have lost by examining contours and glacial elevation change.

## WHAT THE RESEARCH TELLS US

Up until now most North American attention has been focused on the rapid loss of glacier ice in Glacier National Park in the US, where 113 of the 150 glaciers that existed in 1860 have vanished. The reason for that attention was that the Americans actually knew what was happening to their ice and knew also how to attract media attention to what the science was telling them. While there has been coordinated research in the Arctic, up until now we had no clear idea of exactly how rapidly glaciers in Western Canada were disappearing. Now we do. In the last 20 years in our mountain national parks alone, we have lost as many glaciers as existed a century ago in Glacier National Park.

The aggregate loss of ice suggested by the collective findings of WC$^2$N researchers is noteworthy. Some 150 glaciers may have disappeared in the 65 years between 1920 and 1985. Another 150 glaciers disappeared in the 20 years between 1985 and 2005. Through the efforts of IP3 and WC$^2$N researchers we now know that we lost 300 glaciers in the Canadian Rockies between 1920 and 2005. These losses appear to be accelerating. If these findings are substantiated, we may have to reimagine what the Canadian West will be like in the future.

Maybe we don't need to panic. Glacial melt accounts for only about 1 per cent to 6 per cent of the mean annual natural flow of rivers originating on the eastern slopes of the Rockies. But we might want to be looking carefully at when this melt occurs and how that might relate to changes in the timing and amount of precipitation that a warmer atmosphere might bring. The Bow River basin, for example, covers nearly 40 per cent of the area of Banff National Park. In low-flow years, glacial melt supplies 13 per cent of the summer flows at Banff. In extreme low flows, up to 55 per cent of the late summer flow of this river is produced by glacial melt along the Great Divide. Bow Glacier and the icefield from which it flows are receding at startling rates. And though we act as if we are the first to discover this, the fact of rapid glacial recession has been obvious for a very long time.

Early packer and guide Jimmy Simpson, who lived in a cabin on the shore of a lake at the headwaters of the Bow, commented 50 years ago on the rapid recession of the glacier. In an interview with artists Peter and Catharine Whyte on March 30, 1952, Simpson gave the glacier 50 to 100 years before it melted over the horizon. He predicted that other glaciers along the Divide would melt and that Lake Louise and Bow Lake would become sinkholes. He predicted that as a result of glacial melt, the prairies would have trouble with water supply. Simpson and the Whytes concluded that they lived in the best of times and that they wouldn't want to be around to witness the kind of West that would exist after the glaciers disappeared. They are not around anymore, but we are.

## DEFINING THE EXACT AREA OF THE COLUMBIA ICEFIELD

Supervised by Dr. Brian Menounos, Christina Tennant also recalculated the current extent of the Columbia Icefield. In most contemporary literature the area of the Columbia has been estimated at around 325 square kilometres. Based on the area she focused on for her thesis, Tennant estimated the area of the Columbia Icefield to be about 223 square kilometres as of 2005. Depending on exactly which ice masses are technically deemed to be included, the Columbia Icefield may be seen to have lost as much as a third of the area ascribed to it in the 1990s. As the illustration on page 139 shows, portions of the icefield have also diminished considerably in depth. Further research is underway at the Columbia Icefield to more accurately determine both its exact area and its volume.

## WHAT IS THE FATE OF OUR GLACIERS?

The realization that some 300 glaciers disappeared along the Great Divide of the Canadian Rockies between 1920 and 1985 did not come as a surprise to researchers in either the IP3 or the WC²N program. It is not their job to be surprised. Their immediate concern related to ensuring that the analyses were properly conducted and that the findings would stand up to further scientific scrutiny.

In the long-standing tradition of the scientific method, it was important that these findings become a sound foundation for further research. Dr. Shawn Marshall's work on the state and future of Alberta's glaciers demonstrated how science builds upon itself to address serious questions such as those related to the effect climate may have on future water supply in the Canadian West.

In many ways, Dr. Marshall and his colleagues at the University of Calgary are a culmination of what had been collectively gained by both research networks. Marshall's work builds on that already presented by Brian Menounos, Roger Wheate, Matt Beedle and Tobias Bolch, who compared glacial extents in the Western Cordilleran Region between 1985 and 2005 as revealed by Landsat analysis of glacial area. Their work, in tandem with research conducted by Mike Demuth of the Geological Survey of Canada and Garry Clarke

*Dr. Shawn Marshall.*

Photograph by R.W. Sandford

and Joe Shea at the University of British Columbia, indicated that the total area of glacier cover in the nine regions of the cordillera, including the wilderness areas outside the mountain national parks diminished by a total of 3057 square kilometres between 1985 and 2005 (see maps on page 144).

Building on this broader analysis, Marshall's work focused on the outlook for Alberta's glaciers in the future. Tobias Bolch and his colleagues at the University of Northern British Columbia had also expressed changes in glacial cover as percentages of what existed at the time of the last glacier survey, in 1985. What they discovered was that the area covered by glaciers in British Columbia had diminished by around 11 per cent during that 20-year period between 1985 and 2005. What is alarming, however, is that during the same interval, the area covered by glaciers on the eastern slopes of the Rockies in Alberta had diminished by 25 per cent. In other words, in only 20 years Alberta had lost fully a quarter of its glaciated area.

While glaciologists have yet to come out and say it, these findings are a major blow to the established myth of limitless abundance of water in Canada. We may have 20 per cent of the world's freshwater resources, but much of that is water left "on deposit in the bank" after the last ice age. That ice, as $WC^2N$ has clearly demonstrated, is disappearing quickly. Global warming is causing a cryospheric meltdown that resembles in many ways our recent economic collapse. The amount of water that is left in our account in the glacier bank is much less than we expected. Unaccounted-for greenhouse gases are eroding the principal, interest rates are dropping and the amount is becoming smaller all the time. The disappearance of the major glacial masses in the Canadian Rockies will mean there will be less water in our rivers in late summer throughout the West. What we do not know, however, is how much ice is now buried under collapsed moraines or entrained under debris.

Research conducted by glaciologists in the Western Canadian Cryospheric Network suggests that these glaciers are on the way out and that the pulse we expected as a result of rapid warming has come and gone. As renowned hydrologist John Pomeroy has said, we once held that glaciers were canaries in the climate change coal mine. If what glaciologists are saying is true, the canary may be singing its last song. As we shall see, however, this does not suggest that glaciers are no longer important.

Glaciologists painstakingly counted and measured all the visible glaciers that existed in the Alberta Rockies in 1985, and in 2005 they compared those numbers to what the latest Landsat satellite survey revealed. What they discovered was that in 1985 glaciers covered 1053 square kilometres of the Alberta mountains, but by 2005 that extent was only 791 square kilometres, a decrease of 262 square kilometres, or 25 per cent in only 20 years.

The next step in research aimed at assessing the future of Alberta's glaciers involved estimations of the volume of the ice that still remains. Estimating the volume of glaciers is not an easy thing to do. The only reliably accurate method is to use ground-penetrating radar and LiDAR to give a full representation of the depth of the ice and the character of the landscape over which the ice is spread. Such measurements are expensive and were beyond the means of Shawn Marshall's team. Marshall and his colleagues overcame this problem by employing a suite of algorithms to arrive at an average estimate of the volume of ice remaining in the glaciers on the Alberta side of the Rockies.

By utilizing a series of six global, regional and local aggregate scaling parameters, Marshall and his University of Calgary colleague Eric White calculated that the volume of glacier ice remaining in the Alberta Rockies was somewhere between 30 and 110 cubic kilometres. For the purposes of their first-order assessment of the future of glacier ice in Alberta they arrived at an average estimate of 42 cubic kilometres, which became the foundation for rough calculations of how long the remaining ice would last in each of Alberta's mountain river basins, based on climate warming trends witnessed over the past 40 years.

The foundation for these estimates was the mass-balance history of Peyto Glacier, the only glacier in the Rockies for which there is a record of the changes in volume that have occurred in response to rising temperatures over the past four decades. While Peyto Glacier currently covers about 12 square kilometres, it has lost 70 per cent of its volume in the last 100 years. Winter snow depth and duration of cover have been declining since the 1970s (see graphs starting on page 142).

Using a variety of Canadian Global Climate Model scenarios as a means of anticipating the future, Marshall and White projected the mass balance of Peyto Glacier forward to the end of the 21st century. The results revealed a dramatic decline in volume and runoff from the glacier over time.

After projecting the future of Peyto Glacier forward a century, Marshall then extrapolated the results onto surrounding glacier-fed river basins so as to predict changes in the volume of glaciers feeding the Bow, Red Deer, North Saskatchewan, Athabasca and Peace rivers. The results suggest dramatic loss of glacial ice at the headwaters of each of these systems (see graphs on page 143).

Marshall then went on to translate the loss of glacial mass at the headwaters of each of these important river systems into impacts on streamflow over the coming century. Consistent with the projected loss of glacial ice, these projections indicated a substantial negative effect on streamflows over time.

Dr. Marshall noted that current glacial contribution to Alberta's mountain rivers was in the order of 1.2 cubic kilometres a year. He estimated that as this century progresses this volume will be reduced to as little as 0.66 cubic kilometres a year, or about half. This does not mean, however, that glaciers will lose their importance to the hydrology of the West. In parts of Alberta, existing water resources are fully allocated if not over-allocated. Further reductions in flows that will result from the loss of our mountain glaciers will be noticed by those to whom reliable supplies of water are suddenly no longer available.

But given current trends, even the 0.66 cubic kilometres contributed by much-diminished glaciers could ultimately be reduced to zero over time. Marshall reasoned that if in fact the volume of existing glaciers was around 45 cubic kilometres as estimated by his primitive methods – and if current temperature trends persisted – then Bow Glacier would completely disappear in 53 years, or somewhere around 2060; the ice at the headwaters of the North Saskatchewan would disappear in 72 years, or around 2070; and the glacial sources of the Athabasca would disappear in 83 years, or around 2080. Only the glaciers at the headwaters of the Peace and the Red Deer would survive into the next century. Because they are located in high, north-facing cirques, the glaciers at the headwaters of the Peace might survive 97 years, and those in the high mountains at the headwaters of the Red Deer might last 132 years.

Marshall noted, however, that the greatest uncertainty is the accuracy of present-day ice-volume estimates. As mentioned earlier, his research team estimated the volume to be somewhere between 30 and 110 cubic kilometres. Marshall is concerned that his average estimate of between 40 and 50 cubic kilometres may be too low. He went on to say, however, that the volume of glacier ice in the Alberta Rockies was indeed 40 to 50 cubic kilometres at present and that his forecast is for reductions to as little as 5 to 10 cubic kilometres by 2100.

This would represent a 90 per cent loss of the volume of ice currently present in Alberta's Rockies.

But Dr. Marshall also had a number of reservations concerning his findings. He made it very clear to his colleagues that these estimates were based only on "first-order" calculations that have not been verified by more accurate measures of the volume of ice that actually exists in Alberta's Rockies. Such more accurate numbers –including far better estimates of how much ice remained concealed under moraines or debris-covered surfaces – would be needed to test the validity of Marshall's projections of the future state of Alberta's glaciers.

Marshall also voiced his concern about the potential for accelerated glacier loss beyond the rates he employed to make his projections. Consistent with what Christina Tennant had discovered, Marshall too found that most of the glaciers that still remain in the Alberta Rockies are found between 2400 and 2800 metres in altitude. Marshall pointed out that his work on Haig Glacier in the Kananaskis clearly indicated that upward-advancing summer melt has now reached the 3000-metre mark, which is above the altitude at which the most ice still remains. This suggests that if current trends persist, we could witness an even greater acceleration of the loss of glacial ice in the mountain West. From this, Marshall surmised that if the warming trends witnessed since 1985 continue, the glaciers of the mountain West may be on their way out (see photo on top of page 161).

The decline of streamflows in Alberta's major rivers has clear public-policy implications, especially where streamflow is already fully allocated at its current level. The volume of water available to the city of Calgary and to downstream irrigators and prairie communities will diminish. There will be less water available for petroleum upgrading and industrial and other activities in Edmonton. There will be less water available for oil sands activities. There will be less water in both the Peace and the Athabasca systems, which will have implications for aquatic ecosystems in the Peace–Athabasca delta and downstream in the Slave and Mackenzie systems.

It is important to note again that glacial ice is not just water in the bank, so to speak. It is also a refrigerant that moderates climate and slows climate change effects. If our mountain glaciers disappear, the heat presently withdrawn from the atmosphere to melt glacial ice will suddenly be available to further heat the atmosphere. A great many mountains that once were covered in snow and ice are already darkening and drying. Such mountains reflect less light and retain more heat.

What is happening to the world's glaciers is of great importance to the future climate of the mountain West. We already know that long before global warming has finished reducing the length and depth of our glaciers it will already be after our mountain snowpacks, with huge potential impact on everyone who lives downstream.

Changes in snowpack, however, are just the beginning of changes that will converge upon us over time. Ecosystems will change in response to rising temperatures and to longer, hotter summers and reduced late-season streamflows. In anticipating these changes, however, the past may not be a reliable guide to the future.

## PEERING INTO THE FUTURE OF THE MOUNTAIN WEST: ST. ELIAS MOUNTAINS, YUKON

Garry Clarke is one of North America's most respected glaciologists. He was elected a fellow of the Royal Society of Canada in 1989 and served a term as president of the International Glaciology Society between 1990 and 1993. He was president of the Canadian Geophysical Union between 1993 and 1995, and in 2001 he won a Canada Council Killam Research Fellowship.

Dr. Clarke won the Canadian Geophysical Union's Tuzo Wilson Medal in 2003 and was made an Honorary Member of the International Glaciological Society in 2008.

Throughout his long career, Garry Clarke has devoted his research to the understanding of the physics of glaciers and ice sheets. He is famous for his work in exploring the nature of ice flow instabilities that cause certain modern glaciers to exhibit periods of extremely rapid motion. Clarke is also very interested in how ice flow instabilities may have triggered rapid changes in the global climate during the last Ice Age.

Clarke and his graduate students have been studying Trapridge Glacier in the St. Elias Mountains in the Yukon Territory since 1969. The object of their fieldwork is to monitor the glacier as it passes through a complete surge cycle, and from these observations to determine the trigger mechanisms that cause such surges. The research entailed installing dense instrumentation of the glacier bed to permit year-round observation of the mechanical and hydrological processes that are active where the ice contacts the rock surface over which it flows. With more than 20 data loggers in place and more than 200 sensors in continuous operation connected to satellite telemetry, little of the activity of Trapridge Glacier goes unrecorded. Because of the pioneering nature of this work, Clarke and his colleagues had to invent and then construct many of the instruments they use.

A complementary element of Clarke's research has focused on using improved knowledge of ice-sheet physics to construct computational models that simulate the dynamics of both ancient and modern ice sheets. It is in this research that Clarke interfaces with both IP3 and WC²N. Working closely with post-doctoral fellows Faron Anslow, Alexander Jarosch and Christian Reuten, who are experts in modelling oceanic and atmospheric dynamics, Clarke is attempting to unravel

*Glaciologist Garry Clarke in front of Saskatchewan Glacier at the Columbia Icefield.*

Courtesy of Dr. Garry Clarke, WC²N

the secrets of the last Ice Age so as to discover what factors account for the rapid and, for humans, very dangerous surprises that characterized the Ice Age climate system.

As a result of a long and successful career in these research domains, Clarke has also become an expert in determining ways in which subglacial topography can be mapped so that the volume of any given glacier can be determined. He has been working for decades to develop a suite of mathematical tools that will allow researchers to determine the total ice volume of North American glaciers. The great value of this work resides in the fact that the tools Clarke has developed will help us understand how much ice we actually have, how much water is in that ice, how long the ice can be expected to last under a variety of climate and sea level scenarios and what the impacts of glacial recession may be over time. Clarke's work also allows us to begin to imagine what the Western Cordillera might look like in the absence of ice (see images on page 145).

The comparative images Clarke and his modelling team have created appear very innocent at first glance. One image set Clarke presented at the IP3–WC²N conference in Lake Louise in 2009 illustrated the extent of glacier ice that presently exists in the St. Elias Mountains, which, because of their proximity to the Pacific Ocean and the heavy precipitation this brings, is one of the most heavily glaciated regions on earth. Glaciers of 100 kilometres in length are not uncommon here. This is also where rapid changes in the rate of flow of some of the glaciers are most dramatic, where "galloping glaciers" 35 kilometres long suddenly begin moving at 12 metres a day. These are what first drew Clarke's attention to the fact that the dynamics within glacial systems change over time.

Then you look at the next image and see what the St. Elias Mountains would look like if suddenly all that ice were to disappear. It would be a very different place. This may seem a rather academic exercise, if only because it would be a very long time before the amount of ice that exists in the St. Elias range might melt. If anything, there is evidence that some glaciers may actually be growing in this region as a result of warming temperatures generating more evaporation from the Pacific Ocean, which is causing greater winter snowfalls in some basins. What is not simply academic about the illustration is that it shows it is possible to make fairly accurate assessments of what glaciated landscapes might be like in the absence of ice, which suggests that it is now possible to determine the volume of that ice and how much heat it would take to melt it over time. This will be particularly important in the context of characterizing water security in dry regions in southern Canada and in other areas where water supply may be at issue because of growing populations and intensifying demands on water availability brought about by industrial or resource extraction activities.

Unlike the glaciers in the St. Elias Mountains, many of the ice masses in the Canadian Arctic are exhibiting signs of rapid and dramatic recession. While most Canadians don't give climate change in the Arctic much thought except in the context of the natural resources and shipping opportunities that might eventually become available, the loss of glacial ice at northern latitudes is symptomatic of other trends related to changing patterns of snowfall and snow cover that are already beginning to have profound effects on the climate of the rest of the continent. While Canadians may choose to ignore climate change impacts on northern regions, they are less likely to do so in cherished mountain landscapes such as those protected in the western mountain national parks.

Because our mythology as a nation was developed around the notion of a country that spanned the North American continent from sea to sea to sea, we forget that the longest coastline in Canada is not on the Atlantic or Pacific. Canada's Arctic coastline is longer than both our other coasts combined. The fact that climate change impacts are already more obvious in the Arctic than anywhere else in Canada suggests that these kinds of changes should also be obvious in mountain regions to the south, which, because of altitude and cooler mean temperatures, are considered to be southward extensions of arctic conditions.

When you think of climate change, think of where the dawn's light first strikes. It strikes the tops of mountains. When you think of climate change, think also of where the sunlight touches the earth most persistently, and that is where 24-hour summer light falls on the poles. We should be looking to these places first for the impacts of change. A temperature increase of between 1 and 6°C will cause changes in the Western Cordillera. In tandem with the rest of the IP3 and $WC^2N$ research networks, Garry Clarke's work demonstrates that those impacts are already occurring and have been for some time. According to the findings of the $WC^2N$, those impacts will accelerate in this century to such an extent that the landscapes of the Rocky Mountain national parks that Canadians know so well and so deeply love will ultimately be unrecognizable within only a few decades. Though the mountains themselves will look largely the same, much of the ice will be gone and so will much of the water.

## A PICTURESQUE DRIVE TO THE TIME OF ICE

The time-lapse illustrations Clarke and company developed represent the area of Banff and Jasper National Parks roughly encompassing the route of the Icefields Parkway between Lake Louise and the Columbia Icefield. This is arguably the most famous drive in Canada and one of the world's most acclaimed tourism destinations. In the first image we see the glacial landscape in this region as it was in 2002.

If you had been fortunate enough to drive that section of the Parkway on your way to Athabasca Glacier in Jasper in late September that year, the previous winter's snow would be gone from the peaks, and the icefields and glaciers you would have glimpsed would have been of the dimensions depicted in Clarke's three-dimensional orthographic image.

After leaving the Trans-Canada Highway at its junction with the Icefields Parkway at Lake Louise you may have noticed the dazzling whiteness of Macdonald Glacier on Mount Temple and looked with awe at the glaciers gleaming down at you from the mountains of the Lake Louise group, all reflected in the still waters of Herbert Lake. It may have given you satisfaction to know that these high glaciers were likely to persist long into the future, if for no other reason than because of their altitude on north-facing aspects.

As you drove north, you may also have been excited with anticipation at knowing you were going to see many of the 200 other glaciers you were told existed along the route you were taking to Jasper. You may have been very excited about stopping for a night at the Columbia Icefield, where you looked forward to exploring what others have described as one of the most remarkable natural features in the world. There may also have been waterfalls you wanted to see, and sparkling mountain lakes.

You may have noticed that the mountains looked almost surreal in the early morning sunshine. If so, you may have congratulated yourself on waiting until September to travel so that you could experience the rich, oblique, almost liquid light and the electric blue skies that make autumn in the Rockies so famous.

You likely would also have noticed, perhaps with disappointment, that for the first 20 kilometres there were stunning mountains to see but the ice above seemed to be hiding from you. The Waputik Range is a high wall of mountains that gets its name from the Stoney word for mountain goat. The map you were using may likely have indicated the presence of a huge icefield on the top of this range, but you observed that you could only see its rim from the road. The first really conspicuous glacier you passed was the Crowfoot. Named early in the 20th century for its perfect resemblance to the three toes of a frozen bird's foot, it was so close to you it may have seemed as if you could reach out and touch the wrinkled blue ice. But you quickly realized that in a landscape of this scale it is easy to be deceived. An entire valley and the first stirrings of the Bow River separate you from that ice.

As you drive north and look back, you may have noticed

something else. There is an icefield above Crowfoot Glacier so dazzling in the morning light that it's hard to look at. You realize that the valley is a deep trough and that the really substantial ice is cradled in snow-filled basins that are part of a separate world that exists high above the parkway.

You get a much better glimpse of that world when Bow Lake comes into view. The view in 2002 was so stunning you may have had trouble keeping your eyes on the road. If you were travelling early enough in the day, breezes have not yet disturbed the smoothness of the reflection of Bow Glacier and the Wapta Icefield that forms it, clear and brilliant on the horizon. You would not have failed to notice an enormous waterfall pouring down from the melting ice. At that time, this falls was at the headwaters of the Bow, one of Alberta's most important rivers.

You may have read before you took this drive that even though the Rocky Mountains contribute generously to the water supply of the prairie lowlands, all the rivers in the dry south of this province from the Montana to the Red Deer are fully allocated. The Bow River is, by western Canadian standards at least, a heavily utilized watercourse.

The Bow is 657 kilometres long. From its headwaters at Bow Lake to where it joins the Oldman to form the South Saskatchewan, it falls some 1260 metres. The Bow drains about 25,000 square kilometres, or about 4 per cent of Alberta's total land area. It provides 3 per cent of the water that flows on the surface of the province, but in so doing it supplies 33 per cent of the province's population. Though it may not appear so when you look at it, especially where it is protected inside Banff National Park, the Bow was already in trouble even when you saw it in 2002. There are 15 major dams or weirs on the river and its tributaries. Some 45 per cent of Alberta's irrigated land is in this basin. Governments, industry and agriculture were wrestling even then with how to balance already established demand, population growth, expanding oil industry needs and new realizations about the crucial importance of environmental services provided by aquatic ecosystems. Albertans even then were also trying to manage emerging concerns about how to balance reduced flows and increased climate-change impacts against a frontier sense of optimism and abundance. Fortunately, national parks still protected upland watersheds, which would turn out to mean a great deal to the future of the province's water supply.

From Bow Lake north, you noticed that the number of spectacular nature features seemed to multiply almost beyond comprehension. It is only a few kilometres from Bow Lake to the summit of Bow Pass, where, if you were travelling on that perfect September day in 2002, you would have stopped to look down on Peyto Lake and the valley of the Mistaya River. Like you, the other visitors standing on the platform above the lake were likely speechless with awe.

Above you, flowing down from the same icefield that creates Bow Glacier, was Peyto Glacier. It appeared enormous beyond comprehension. The sound of rushing streams rose and fell with the wind. Below the glacier was an outwash plain where the primordial Mistaya River braided itself together to pour into Peyto Lake. Peyto Lake was a colour of turquoise you had never seen before in nature. What you were looking at seemed less like a lake than a light. Filled with fine, glacially derived sediments, the water absorbs everything in the visible-light spectrum except a shocking blue, which seems to radiate out of the water to make the lake appear faintly neon.

In crossing Bow Pass, you saw from the map that you had passed from the headwaters of the Bow River to those of the North Saskatchewan. The Mistaya, which flows north from Peyto Lake, is a tributary of this great Canadian river. From the viewpoint above Peyto Lake you could see the entire Mistaya valley and count the lakes that have been formed by the damming of spring snowmelt which were maintained in late summer by the meltwaters of the rapidly receding Peyto Glacier.

You may have marvelled at the grand scenes that come into view as you make the long descent down the north side of Bow Pass into the Mistaya valley. Mistaya is a Stoney word for grizzly bear and this valley is one of the great bear's haunts. You noticed also that there was a lot of ice on the rim high above you that comprises the peaks forming the Great Divide, the stone spine that separates Alberta from British Columbia. You may have wondered what it would be like to be up there to see for yourself how much ice there really was. You may have also noticed enormous peaks towering above the Waterfowl Lakes as you plunged downvalley, following the course of the Mistaya to where it joins the nascent North Saskatchewan.

The map told you that behind those peaks, in the Howse valley, was one of the largest remaining icefields in the Canadian Rockies. You couldn't see the Freshfield Icefield from the Icefields Parkway. It was one valley over. But at Saskatchewan Crossing you could look up the Glacier River valley and see enormous glaciers at its distant head. What you were seeing were the Lyell and the Mons icefields, two of the lesser-known of the 12 major icefields encompassed by the Canadian Rocky Mountain Parks World Heritage Site. On that perfect day in 2002 you remember that the glaciers pouring down from peaks named after early Swiss mountain guides looked as though they are near at hand but you knew they were not. Most people took two days to hike to the Lyell.

There was, however, plenty of ice to beckon you northward on the Icefields Parkway. You saw the lip of a huge glacier on the back of Mount Wilson just above Saskatchewan River Crossing. As you rose upward from the lower valley through which the North Saskatchewan quietly makes its way out of Banff National Park on its way to Edmonton and ultimately Hudson Bay, you may have noticed more ice along the Great Divide. There were huge glaciers on Mount Amery and ice seemed to cascade down every shoulder of Mount Saskatchewan. Then, suddenly, at the top of a big hill leading to a loop the highway makes on itself over a wide gravel flat, you saw the glaciers on Mount Athabasca and the lip of the Columbia Icefield itself. The landscape seemed so monumental in scale, however, that it was hard to comprehend. It stunned you that the road kept going up and up. It stunned you, too, to see how many waterfalls there were and how much water was still falling even in September.

At Parker Ridge you may have gotten out of your car to stretch your legs. Here you may have found your mind slowing with the scale of place and intimations of time operating on a geological scale you had never experienced before. You may have walked the two kilometres up the ridge itself into the wind and sun to view the biggest glacier of them all, the ten-kilometre-long Saskatchewan, surreally flowing from the great icefield that formed it. By any standard, the Columbia Icefield is a stunning geographical feature. It is a high basin of accumulated snow and ice that straddles what you were told then was an area of 325 square kilometres, roughly 125 square miles, along the Great Divide. Located at 52° north latitude and 117° west longitude, it was held to be the largest concentration of glacial ice in North America below the Arctic Circle. It is one of the last places in southern Canada where cold, wind, weather and water still interact in exactly the same way they did during the last Ice Age. What you marvelled at was its accessibility.

At Athabasca Glacier you discovered that you can literally get out of your car and in a few moments walk directly back into the Pleistocene, a colder epoch in the earth's history when much of North America was buried beneath a two-kilometre carapace of ice.

Here one is immediately stood upright by the cold wind blowing off the deep ice. Snow that fell and was compressed into ice 400 years ago is melting at the glacier's snout. The towering black peaks seemed to lean over you. The familiar sun was a cold and distant star. You remember that there was a different sense of time here. Fleeting hours hardly mattered. The day seemed the smallest unit, the season next, then the year. Beyond the year there was only the timelessness of epochs, the incomprehensibly vast passing of the geological seasons, mountains rising and falling and the coming and going of entire rafts of planetary life. The grumbling ice may have humbled you. Confronted by eternity, you may have felt small and irrelevant. It may have occurred to you that epiphany was possible here, a sense of aesthetic arrest. A shudder may have run through your soul as you realized, suddenly, what an Ice Age really is and what ice means.

Beyond the experience of the utterly monumental in nature, you may have returned home changed. Even a brief visit to the Columbia Icefield teaches us that water is a central element in defining what the surface of the earth will be like at any given moment in its history. A change of only a few degrees in atmospheric temperature defines what forms water will take and how it will act in shaping the world and the life forms that exist upon it at any moment in earth time. Of all the extraordinary glacial features offering testimony to the importance of water in the making of the world that exist at the Columbia Icefield, Snow Dome is perhaps the most amazing, for it is a triple continental divide.

## WHY SNOW DOME IS IMPORTANT ...

A divide is the boundary of a water drainage basin. A continental divide is the boundary between water that flows into two different oceans. A triple continental divide is rare in nature. It is the uppermost point that separates water that flows in different directions across a continent to pour ultimately into not two but three separate oceans. The apex of the triple divide in the Columbia Icefield is the summit of Snow Dome, a 3451-metre peak that overlooks Athabasca Glacier from the north. Snow falling on the summit of this mountain can, depending on where it falls, end up in any one of three major river systems, each bound for a different sea. From the gently dipping northern shoulder of the mountain, melt joins the waters of the Athabasca River, which flow nearly 3000 kilometres northward through the massive Mackenzie River system to become part of the Arctic Ocean. Meltwaters from the mountain's western shoulder flow through the Bush and Wood rivers into the Columbia, which meanders southward for 2000 kilometres to join the Pacific near Portland, Oregon. Meltwaters originating from Snow Dome's eastern shoulders flow into the North Saskatchewan, the great stream of the northern Canadian plains. This river flows nearly 2000 kilometres to pour into the Atlantic Ocean at Hudson Bay.

While Snow Dome was revered initially as a worthy mountaineering objective and then as an interesting hydrological feature, it has come to stand for a great deal more as we continue to expand our understanding of the importance of the Columbia Icefield as a climatic thermostat and water tower for western North America. If you are among those who believe that life itself has the collective capacity to self-regulate the surface temperatures of earth so as to perpetuate conditions ideally hospitable to the sustained optimization of life, you will be interested to know that the conditions on Snow Dome also represent "the triple point of water." The triple point of water is the temperature range at which all three phases of water – solid, liquid and vapour – coexist and interact. This range, which is absolutely crucial to tolerable climate variability and to life processes, resides between -40°C and +40°C. At the centre of this range is the freezing point. Looking up at the lip of ice visible from the Icefields Parkway, one gets a sense of what this means to the landscape and those who live upon it. Our entire existence is clustered around the triple point of water.

The triple continental divide on Snow Dome is unique also in that it is so deeply blanketed in glacial ice. Over the last two decades the glaciers of the mountain West have been shrinking faster than at any other time in recorded history. If the world continues to warm and all this "water in the bank" disappears, the West, and our entire continent, will be a very different place.

In this larger context, Snow Dome could be seen as a triple triple divide. Not only does it separate mountain meltwaters to send them on their way to three different oceans, it does two other things also. It reflects the atmospheric temperature balance that determines the proportion of water that exists on the planet in each of its various and highly active forms; and it functions as a portent, dividing past hydrological regimes from those of the present and those that will exist in the future, not just in the Rockies but in the entire West. The scientific research outcomes of the IP3 and WC²N put this fact into even more disturbing relief. Examining the time-lapse animations created by Garry Clarke and his colleagues, we can visualize what kind of West is likely to exist if the warming trends we have seen over the past 85 years persist. If the characterization of the changes we might expect are even partially accurate, Banff and Jasper national parks are not going to be even remotely like they are now. If that is the case, one wonders what the rest of the West will be like.

## ... AND HOW THINGS ARE CHANGING IN THE MOUNTAINS

Let's project the experience that would be available to a traveller who might visit the landscape encompassed by Clarke's time-lapse map in 2100 and compare that to what exists now. Travelling by whatever means will be available 85 years hence, the experience would not resemble anything we would have had in 2002. Starting up the Icefields Parkway from just north of Lake Louise, the first thing a visitor would be grateful to notice is that there were still glaciers on the high peaks surrounding Moraine Lake and Lake Louise. North-facing and high in altitude, and in some cases also protected by collapsing moraines, these glaciers had yet to see their ice melted away by the dramatic warming in the valleys and the longer summers. You do notice, however, that there is no longer any perfect reflection on Herbert Lake to mirror their glory. There is no reflection at all, because there is no lake. There is no lake because there is no water. There is less water anywhere around here, because there is less snow and it melts earlier. A hundred of the 395 named lakes that existed a century ago are no longer there, though a few new ones have come into existence as a result of the rapid decline of glacial ice. And there is a lot less ice. You notice this because the lip of ice that marked the edge of the Waputik Icefield has receded beyond view and Crowfoot Glacier is gone altogether, along with the icefield that formed it.

With photographs taken in 2002 as a reference for what existed before, you proceed north on the Icefields Parkway to Bow Lake. You notice that the lake is still there, but what really shocks you is that the glacier that supplied the water that formed the lake is gone. It has been gone for nearly 50 years. While the Bow River still flows out of Bow Lake, its flows are greatly reduced. While the river continues to flow all year round at Calgary because so many other tributaries contribute meltwater and groundwater and dams control the flow, in the high mountains at its source the river dries up in late summer during drought years. Like many waterfalls, it too has become but a freshet.

Even the icefield out of which Bow Glacier once flowed is gone. Most of the waterfalls in the Rockies, you discover, now exist only briefly during the intense melt of an earlier spring, as do many streams and even some rivers. Fewer people now live downstream, you notice, because there is less water than there used to be and flows are less reliable. Though you can see from the map that a tiny remnant of the Wapta Icefield still clings to the highest north-facing peaks, the glaciers that flowed from it are completely gone. This, you realize, must have had devastating consequences on both sides of the Great Divide. You realize that on the west side of the Divide, where the Wapta Icefield once draped over the eastern boundary of Yoho National Park, the effects have been catastrophic. Yoho Glacier has disappeared, as has des Poilus. Daly Glacier has vanished and the loss of meltwater has turned Takakkaw Falls, once one of the great wonders of the world, into a spring freshet. There is no trace of ice anywhere on the Highline and Skyline trails. The age of ice appears to be in full retreat (see images on pages 146–147).

The full impact of what has happened, however, does not make itself evident until you visit Peyto Lake. The glacier is gone and only a thin stream of water pours down from the remnant of the icefield that once formed it. You can no longer hear the sound of mountain water rising and falling in the wind. The lake is smaller by half than it once was, a century earlier, but the difference that bothers you the most is that the lake has lost its amazing colour. The light that was once the lake has gone out. It no longer glows a bright turquoise in contrast to the striking green of the surrounding forest. You realize that few glacial lakes remain that still contain the huge volumes of rock flour that need to pour into them to give them such bright colour. They have gone from a glowing turquoise to a sedate blue. Everywhere, it seems, the Rocky Mountain lakes have lost their lustre. You find yourself speechless with a different kind of awe.

The big glaciers that clung to the Great Divide on the west side of the Mistaya valley are all gone. Snowbird Glacier, once regarded as one of the most exquisite natural features in the Rockies, has disappeared. The lakes that once sparkled in the Mistaya valley are diminished in size and depth or gone. Though you can't see it, because it is a range away, the maps you have with you indicate that a tiny spur of the Freshfield Icefield still survives, but the greatest of that icefield's glaciers, the Freshfield itself, has vanished into thin air.

At Saskatchewan River Crossing you look west in the hope that the Mons and the Lyell icefields have survived. The Mons, however, is completely gone. It is as if it had never been there. Only fragments of the Lyell remain – high, distant patches with no glaciers at all flowing forth. You discover that the Glacier River ceased flowing permanently 50 years ago. Wilson Glacier above Saskatchewan Crossing has shrunk so much it is no longer visible. A few glacial remnants linger on Mount Amery and the highest shoulders of Mount Saskatchewan, but the real shock doesn't come until you reach the Columbia Icefield. There is no longer ice on the horizon at the loop below the Big Hill. Astonished by the late-summer heat, you stand on the crest of Parker Ridge only to discover that Saskatchewan Glacier no longer remotely resembles the Ice Age colossus that poured grandly out of the Columbia Icefield a century before. You realize that the Columbia Icefield could be as much as 200 square kilometres smaller than it was thought to be in 2002 and you suddenly fear what you are going to discover when you drive the eight kilometres between Parker Ridge and the once world-famous Athabasca Glacier.

All your fears are realized when you turn the corner at the site where the Icefield Centre once sat. The Columbia Icefield is no longer one of the last remaining places in southern Canada where cold, wind, weather and water still interact in exactly the same way they did during the last Ice Age. Where a century ago visitors marvelled at the accessibility of the five-kilometre-long glacier, what remains is almost impossible for anyone but a mountaineer to reach. Getting out of your electromagnetically powered vehicle, you no longer walk back into the Pleistocene, that colder epoch in the earth's history when much of North America was buried beneath two kilometres of ice. You walk instead into a climate-change nightmare. It is a nightmare that Garry Clarke and his colleagues fully visualized a century before (see images on pages 148–149).

Like Shawn Marshall, Garry Clarke was highly circumspect about the accuracy of the volumetric analysis that formed the foundation of the rough estimates he and his team put forward regarding how long we can expect glaciers to survive on the landscape if current climate warming trends persist. Clarke pointed out that he and his group had had considerable difficulty estimating the exact present volume of Athabasca Glacier. The problem had to do with how to accurately characterize the underlying bed surface beneath the third icefall on the glacier, which he and other researchers before him had estimated to be at least 300 metres deep. How long that ice would remain during periods of continued warming was hard to determine. Another problem that challenged Clarke and his colleagues was uncertainty about the extent to which the lower surfaces of the glacier might become moraine-covered, which would result in an insulation effect that would allow lingering ice even at lower altitudes to last much longer.

These projections, however, offer important insight into what is happening to water we used to have in the bank in the North and in the mountain West. What is happening to our glaciers is of great importance to the future climate of the mountain West. While glaciers may not be as important in terms of total water supply as they once were, what is happening to glaciers may be a warning of other, far more significant threats.

We already know that long before global warming has finished reducing the length and depth of our glaciers it will already be after our mountain snowpacks, and that could have a huge influence on our water supply.

Reduced flows in western rivers will affect power generation and agricultural, industrial and municipal water security throughout western Canada. These flow reductions will also affect interprovincial water-sharing arrangements and trans-border agreements with the United States such as the Columbia River Treaty. But that is only the beginning of the changes warming will initiate.

The long-term ecological effects of widespread glacial recession and diminution of snowpack and snow cover remain to be seen. We should expect, however, that the reduction in the amount of light reflected by ice and snow in this region will have further warming effects on climate and that these effects will cascade through both terrestrial and aquatic ecosystems.

Beyond creating powerful images of why we need to act on the climate change threat, Garry Clarke and his colleagues in the Western Canadian Cryospheric Network make an articulate and forceful case for why further research, into increasingly accurate modelling parameters that would enable more reliable prediction of future hydrological and climatic states, is crucial to the very survival of our current economic and political structures.

# CHAPTER 3

## *What the Snows Are Telling Us: Marmot Creek and the Changing Hydrology of the Mountain West*

IP3 research was undertaken in nine river basins. These include Marmot Creek in Alberta's Kananaskis Country; the Lake O'Hara basin in Yoho National Park; Wolf Creek in the Yukon; Havipak Creek, Trail Valley Creek, Baker Creek and Scotty Creek, all in the Northwest Territories; Polar Bear Pass in Nunavut; and Reynolds Creek in Idaho in the northern US, plus Peyto Glacier in Banff National Park.

By synthesizing what is learned in each of the IP3 research basins, scientists hope to better understand hydrometeorological processes in Canada's cold regions. Better understanding of these processes will allow researchers to improve their understanding of the parameterizations that characterize processes that are taking place at the local scale in cold regions. These refined parameterizations will enable modellers to make more accurate predictions as they seek to define how specific land-use changes and a globally warming atmosphere will affect where and how we live in each of the major river basins in the mountains of Western Canada and in the Yukon and Northwest Territories.

Accurate parameterization is not easy, however. From *Snow Ecology: An Interdisciplinary Examination of Snow-Covered*

*Ecosystems*, edited by Gerry Jones, John Pomeroy, Skip Walker and Ron Hoham, we learn why. Though water is a very simple compound in a chemical sense, attempts at parameterization demonstrate that water's interactions at every level of the natural world are complex almost beyond comprehension. Above 0°c, water exists as a liquid or as vapour. Everything to do with water changes with the 0°c isotherm, however – that constantly moving line, across a vertical map of the atmosphere, above which water freezes. Water acts one way in its liquid form and completely another way when it takes the form of snow or ice. It acts in yet another way when it becomes a vapour. But it can also exist in all three of these states at once. Below 0°c, obviously, water exists primarily as snow or ice. But even at temperatures colder than the freezing point, some water can still exist as vapour or as a thin, liquid-like layer found on the edges of snow crystals.

Because of the variation in daytime, nighttime and annual temperatures, most of the earth's snow exists only seasonally. Conditions within seasonal snowpacks can vary dramatically, even during the course of a single day. Snow can quickly melt to liquid or sublimate to vapour. The liquid and vapour forms will then interact with remaining snow. The entire snowpack can then refreeze at night.

## MODELLING THE PHYSICS OF SNOW

Hydrological modellers understand that the liquid water content of a given snowpack declines as the temperature drops below 0°c. In terms of parameterization, however, there are other, complicating factors that must be considered. These include the rate of melt within the snowpack, the extent of rainfall on the snow-covered surface, the capacity of the snow to retain liquid, and the rate at which water is able to drain from the pack. The combination of these factors will determine how quickly the liquid water content of the snowpack will change.

Snow also has other, almost otherworldly qualities that need to be understood by modellers. It takes a lot of heat to melt snow. This is called the latent heat of fusion. The amount of energy required to melt 1 kilogram of snow that is already at 0°C can be as much as the amount of energy required to raise the temperature of 1 kilogram of liquid water 79°C. This is why snow is such an effective climatic refrigerant. Fortunately, however, this is a reversible process. Comparable amounts of heat are released into the environment when liquid water freezes.

It also takes a lot of heat to sublimate snow. The amount of heat required to cause sublimation is defined as the latent heat of vaporization. The energy required to vaporize 1 kilogram of snow is roughly equivalent to the amount of heat required to raise the temperature of 10 kilograms of liquid water 67°C. This is why it can be cold in the spring in the presence of snow-drifts even when the sun is warm. Heat is being sucked out of the air by the sublimating snow. Vaporization too is reversible. Huge amounts of energy are released into the environment when water vapour recrystallizes into ice as hoarfrost.

The fact that it takes so much energy to freeze or vaporize water makes it no surprise that the conduction of heat by snow-packs is low when compared to underlying soil surfaces. The thermal conductivity of snow, however, varies with its liquid water content. In many circumstances the thermal conductivity of soil can be six times greater than that of the snow above it. This means that a layer of snow can insulate over six times more effectively than the equivalent depth of soil. The total insulation powers of snow, however, depend upon its depth.

Modellers wishing to forecast the hydrological future of snow-covered landscapes must also take the effects of sunlight on nival systems into account in their predictions. Compared to soil and vegetation, snow reflects a high proportion of shortwave solar radiation. The reflectivity of a given surface is called its albedo. A fresh, continuous accumulation of snow can have an albedo of 80 to 90 per cent of the shortwave radiation that falls on it. It is this high degree of reflection that causes snow blindness and produces sunburn even on winter days. The shortwave radiation that is not reflected is largely absorbed by the upper 30 centimetres of the snowpack. The moment the extent of snow cover diminishes, however, albedo drops. Bare soil and vegetation can absorb as much as eight times as much shortwave radiation as fresh, continuous snow cover.

It is interesting to observe that while snow reflects shortwave radiation, it absorbs longwave radiation. It is rather like a black body in this respect. Longwave, or thermal infrared, radiation that falls on snow is absorbed and reradiated as thermal radiation. The wavelength of such emissions depends on the surface temperature of the snow.

In addition to the parameters associated with the nature of snow itself, modellers also have to contend with the aerodynamic qualities of snowpacks. Land surfaces and vegetation typically offer several orders of magnitude more resistance to wind than a smooth snow surface. As a result, the transfer of heat between the atmosphere and the ground is much greater for vegetated and forested surfaces than for open snowpacks.

If snow cover persists, it becomes a habitat in itself and an ecosystem separate from the landscapes upon which the snow has fallen. Snow is a unique environment with its own remarkable hydrological conditions. All organisms living in or under a given snow ecosystem must contend with these conditions in order to survive and must take advantage of them if they are to enhance their adaptive capacity through natural selection.

Some of the most remarkable adaptations in all of nature occur within snow ecosystems.

Creatures that thrive at or below the freezing point of water are known as cryophiles. They like the cold. The best-known cryophiles living in this area of the Rocky Mountains are the red algae found in snow in concentrations large enough to turn the spring and summer snowpack pink. Snow algae are what are known as "obligate cryophiles" in that they have no choice in where they live: they live in cold environments and nowhere else. Snow is their principal habitat. The ranges of difficult environmental conditions these organisms face are extraordinary indeed. They include huge extremes, not just in temperature but in acidity, radiation levels and nutrient availability. In some cases they also include conditions of extreme desiccation that occur when liquid water is no longer available.

Obligate cryophiles such as red and green snow algae are also called psychrophiles. The psychrophiles are an amazing lot. For many of these organisms, optimal growth can only occur at temperatures below 10°C. Minimum temperature for growth can be 0°C or even lower. The survivability of these algae at extremely cold temperatures is truly amazing. In laboratory tests, viable cells of snow algae have recovered from temperatures as low as −196°C. Even the hardiest psychrophiles, however, need liquid water at some point in their life cycle in order to grow and reproduce. The phases of the life cycle of snow algae correlate with physical and chemical factors at the time of snowmelt. When snow algae bloom, populations of up to one million cells per millilitre are not uncommon.

## MUCH DEPENDS ON COLDNESS

From all this we see that a great deal depends on temperature. Cold matters. Warmer global and regional temperatures will bring about a wide range of climatic changes in snow-covered regions. Because a warmer atmosphere can hold more moisture, we can also expect more overall precipitation during the typical cold season. We can expect changes in snowfall regimes. We can expect more rain and less snow. With warmer temperatures we can also expect more mixed precipitation and more frequent rain-on-snow events. Warmer temperatures will also bring about changes in the properties of snowpacks. Stable snow cover will last for a shorter period of time. We can also expect a decrease in snow depth and a decrease in albedo due to more rapid metamorphosis of snow occasioned by more frequent midwinter thaws and rain-on-snow events. Snowpacks are likely to become denser and icier, which will affect their insulation properties.

Warmer temperatures will also bring about changes in the amount and timing of snowmelt, which will alter the entire seasonal runoff pattern. More meltwater will go to the soil, to be released after the snows have disappeared. As a result, less snowpack will be available for spring snowmelt flooding. As the summer season becomes longer and less snow is on the ground and for a shorter time, there will also be changes in evaporation and soil moisture regimes. It remains to be seen, however, whether a potential increase in precipitation will make up for the loss of soil moisture due to higher temperatures and longer summers.

There will also be a great deal more variability in climatic conditions. The effect of these changes on snow-influenced ecosystems will also be highly variable. While contemporary ecosystems have had to adjust to changing climates in the past, the changes that are taking place today are much more rapid than those that have occurred in the past. Given that, at the time of this writing at least, there has been no effective global or national action on the climate threat, we should expect these rates of change to accelerate throughout the rest of this century.

A change in patterns of snow cover will have a huge effect on the hundreds of organisms that inhabit winter snowpacks. There will likely be dramatic spatial changes in their habitat which may require changes in their seasonal life cycle. Perhaps just as significantly, evidence suggests that other ecological communities are already responding to the warming global climate and to the climate and snow cover feedbacks this warming is causing. These changes and resulting feedbacks represent an enormous challenge to scientific research. As Pavel Groisman and Trevor Davies explain in their chapter of *Snow Ecology: An Interdisciplinary Examination of Snow-Covered Ecosystems,* snow and the living communities that exist within snow interact with one another in ways that we are only beginning to understand. Much remains to be done if we are to be able to parameterize the processes that sustain snow ecosystems and model the effects of these ecosystems on the physical and geochemical properties of snow.

What we do know now is this: seasonal snow cover interacts directly with other elements of the climate system, not only at the interface between the snow surface and the atmosphere but at local, regional and hemispheric levels as well. Seasonal snow cover therefore greatly affects not only the ecosystems that are within or beneath the snow, but through its influence on climate at a hemispheric level, the presence of extensive, long-lasting snow cover can also affect ecosystems far distant from the direct effects of winter cold. The behaviour and future distribution of snow cover is thus of considerable interest to ecologists, because of both its local effect on habitat and its cooling effect on overall atmospheric temperature, which ultimately affects the stability of all other ecosystems on earth.

At present we don't know a great deal about how the potential loss of snow cover will affect the global thermostat. Data relating to the extent of snow cover in many parts of the world extends back a century. There are concerns, however, about the representativeness and reliability of much of this data. As Groisman and Davies explain, satellite observations did not start producing reliable data on the extent of hemispheric snow cover until the 1970s. Over the ensuing 30 years, satellite comparisons show that snow cover has decreased in the northern hemisphere by some 10 per cent, with even greater decreases in the summer months, a finding that suggests that the same warming that is causing rapid and widespread glacial recession has also begun to reduce snowpack and snow cover. The prospect is for continuing diminishment of both the depth of snowpacks and the area of snow cover in the coming decades. The increasing loss of snow cover is bound over time to affect ecosystems. It is anticipated that these changes will accelerate warming through feedbacks generated by changes in surface albedo and increased temperature at the interface between what were once seasonally snow-covered landscapes and the atmosphere. The fear is that the large-scale changes in the depth and extent of seasonal snow cover that we are already witnessing will alter atmospheric circulation patterns and further intensify the global hydrological cycle.

The exact nature of the feedback processes between snow cover and local and global climate systems is poorly understood. In order for humanity to adapt to the changes we are causing in our climate, it is important that we learn as much as we can about snow cover processes and incorporate this knowledge into predictive models.

## SNOW MATTERS

John Pomeroy knows at lot about snow. Dr. Pomeroy has

conducted research in western and northern Canada, the United States, Bolivia, Russia, Wales, Scotland, Nepal and Japan and authored over 200 research articles and several books. He serves as president of the International Commission for Snow & Ice Hydrology, and besides being the project lead for the IP3 Network, he is also a member of the executive committee of the Global Institute for Water Security at the University of Saskatchewan. Though Dr. Pomeroy is involved in the administration of a great number of research projects, his first love is fieldwork. His ongoing research interests relate to snow physics, mountain hydrology, hydrological predictions in basins where there are no streamflow or other gauges, and the impacts of land use and climate change on the hydrology of cold and semi-arid regions.

Years of studying snow-covered landscapes all around the world have given John Pomeroy some very interesting insights on how snow accumulates and distributes itself, particularly in mountain landscapes. He has observed, for example, that patterns of snow distribution are different depending on the scale at which they occur. At the very largest, or macro, scale, snow distribution is affected by planetary-scale meteorological effects. These are colossal forces. The oceans of the world and our atmosphere form a continuum in which the one is constantly influencing and affecting the other. In tandem with planetary life, our atmosphere is self-regulating in terms of temperature; self-protecting against radiation from space; and self-perpetuating in terms of composition. If we look skyward we are confronted with marvel after everyday marvel. We forget the power of these systems. It is at the interface between sea and sky that hurricanes 750 kilometres in diameter swirl air that weighs 8.8 million tonnes per square kilometre around a still eye at 300 kilometres an hour. It is from this interface that 28,000°C, 150,000-volt,

435,000-kilometre-an-hour lightning bolts flash from the sky to strike the earth right before our very eyes. It should be no surprise that such an active and powerful place as the global atmosphere should influence snow distribution at the macro scale, but how it does so is also interesting.

Snow distribution is affected at the planetary level by atmospheric flow deviations such as the Coriolis effect. The Coriolis force is created by the world spinning within the loose blanket of its own atmosphere. It is the effect of the earth's rotation that deflects a body of fluid or gas relative to the earth's surface to the right in the northern hemisphere and to the left in the southern hemisphere. The Coriolis effect is at its maximum at high latitudes and zero at the equator. At the macro scale, snow distribution is also affected by standing waves in the atmosphere, the flow around mountain ranges, latitudinal temperature changes and the proximity of regional moisture sources such as oceans. At the middle, or meso, scale – that is to say at distances of 1 to 100 kilometres – snow distribution will be affected by local and regional changes in topography and the convective influence of precipitation originating from the addition of heat and water vapour from lakes. At the smallest, or micro, scale, the effects of macro- and meso-scale influences are largely associated with redistribution of snow after snowfall. These effects include the movement of snow by wind and the effects of settling and metamorphosis of the snow at it ages.

As a result of these different effects occurring at different scales, different places have very different kinds of snow and very different snowfall patterns. Tundra and prairie snows are different from those in taiga or boreal regions. Alpine and maritime snows are the deepest and the most complex in terms of structure. These snows are composed of different layers of varying grain size and wetness and are characterized by the presence of ice bodies within them that form as a result of surface melt permeating the snowpack. Ephemeral snow has its own unique properties.

In the Great Lakes region of North America the fact that enormous bodies of water remain open throughout much of the winter dominates both the character of the snow and the patterns in which it falls. The Great Lakes are the engines driving regional snow distribution. Huge volumes of water vapour are continuously available for the production of snow. As topographical influences are comparatively minor in the Great Lakes region, the major factors in snowfall distribution include the nature and character of winter storm tracks, the extent to which temperature decreases with latitude, and how much cooling occurs in air masses that move from water to land. Upward convection of air occurs over relatively warm lake water, producing cloud and then snowfall, which moves inland on prevailing winds. These relatively warm air masses rise when they reach the lakeshore, cool, and then produce sometimes very large volumes of snow.

The importance of this effect can be seen in the relative amounts of snowfall that occur in southern Ontario. Few Great Lakes storm tracks reach southwestern Ontario, where average winter snowfalls are around 100 centimetres. Around Lake Huron, however, where there is a prevailing northwesterly flow of air over the lake, average snowfall is around 320 centimetres. Having grown up in Cleveland, Ohio, John Pomeroy has first-hand experience of the famous snowbelt that extends from Cleveland to Buffalo as a result of convection storms arising out of evaporation from Lake Erie and Lake Ontario. Because of the amount of moisture available and the temperatures at which it forms, snow in the Great Lakes region is often wet and heavy, very different from the snow that typically falls in the Rocky Mountains.

The depth of snow cover in mountainous regions increases with elevation because of the increasing number of snowfall events that take place there and because melt is slowed by the cooler temperatures that usually accompany high altitudes. In the mountains of western Canada, the influence of elevation on the extent and nature of snowfall is most pronounced at elevations above 600 metres. It has also been observed that the increase in snowfall in association with elevation is most pronounced where moisture-laden wind-flows ascend mountains rather than where descending flows occur. This effect is very pronounced in the Selkirk Range for example, where enormous snowfalls occur regularly on windward slopes, making places like Rogers Pass famous for snowfalls of several metres that lead to incredible winter snow accumulations. Snowfall is less on the lee side of mountain ranges. This is particular evident in the Rockies, where most of the moisture carried aloft from the Pacific Ocean has been largely deposited as snow before prevailing winds arrive at the Continental Divide. This too affects the character of the snow that falls. The snow in Banff National Park is very different from that in the Great Lakes region. It is light and dry, often so light and dry that it is like powder, perfect for skiing.

Elevation, however, is not the only influence on how snow cover is ultimately distributed over mountain landscapes. Other factors include slope, aspect, the nature and extent of vegetation, wind direction and speed, temperature and prevailing weather patterns. Wind and slope are important factors in determining both the extent and distribution of snowfall. Studies in the Colorado Rockies have shown that long-term average snowfall is strongly linked in some places to topography located up to 20 kilometres upwind. The same research demonstrated that long-term average precipitation in that area of the Rockies was not linked specifically to altitude except at points along the same mountain ridge. Pomeroy and his colleagues have found this to be the case also in the Kananaskis region of the Canadian Rockies (see diagram on page 150).

## THE ULTIMATE SNOWBLOWER

John Pomeroy and others have also discovered that wind has an enormous impact on both the character and deposition of snow. Over 25 years ago Pomeroy and engineer Tom Brown developed an opto-electronic blowing-snow-particle counter which permitted detailed measurements during snowstorms. This particle-counting technology has been used in the Prairies, the Arctic and the Antarctic as well as in mountains in the US, Canada and Scotland. With this device, Pomeroy found that blowing-snow transport involves three different kinds of movement. "Creep" is the term for the movement of particles that are too heavy to be lifted by the wind but light enough to roll. "Saltation" is the movement of snow crystals by way of a skipping or jumping action along the snow surface. "Turbulent diffusion" is the horizontal movement of snow particles in suspended flow close to the interface between the snow cover and the moving air. Suspended snow bouncing along the snow surface can be lifted from the top of the saltation layer by turbulent eddies in the atmosphere where the wind can carry it hundreds of metres above the ground. Pomeroy and others have observed that the concentration of suspended snow above continuous snow cover reaches a maximum just above the layer of saltation activity and decreases with height at a rate that depends on wind speed.

The threshold wind speeds required to produce each of these forms of movement in a variety of different snowpack conditions and temperatures are now known. In northern prairie environments, Pomeroy found that it only takes a wind speed of 7 metres per second, or just over 25 kilometres per hour, to begin to move dry snow at −25°C. At 0°C, however, the same snow won't begin to move until the wind speed reaches 9.4 metres per second, or 34 kilometres an hour.

The greatest concentration of snow transport normally takes place at between 10 and 25 centimetres above the snow surface. In one study, Pomeroy and his colleagues demonstrated that 77 per cent of the movement of snow through saltation takes places below a height of 1 metre at a wind speed of 10 metres per second, or roughly 36 kilometres an hour. Big storms move a lot of snow around. It was seen that some 40 per cent of total snow cover could be transported at wind speeds of 30 metres per second, or about 108 kilometres an hour. Such wind speeds are not uncommon on the Canadian prairies and are likely to be more common as the global atmosphere becomes more energetic because of higher mean temperatures. Once again, cold matters.

In 1990 Pomeroy and his mentor Don Gray were able to model saltation transport in relation to threshold wind speeds and the influence of vegetation. They found that although saltation transport starts at lower wind speeds over fresh, loose snow, not a great deal of snow is transported at lower wind speeds. The scientists were surprised to discover that the real action begins at higher wind speeds over wind-hardened snow. They also observed that the effects of wind on the evolution of snow cover are most pronounced in open environments. They found that exposed vegetation rapidly diminishes the amount of wind energy available to transport snow, and that little or no transport occurs within stands of dense vegetation until the snow depth is within a few centimetres of the vegetation height. This finding is consistent with what other researchers would later discover in a variety of IP3 research projects in the Canadian Arctic.

Wind distribution of snow results in the erosion of snow

cover by the sheer physical force of the wind. It breaks the bonds that hold snow crystals together and overwhelms the cohesive forces that wetness creates in the snowpack. Wind literally breaks snow apart and granulates it. As a result, the density of snow redeposited by wind can be up to six times greater than that of snow that has fallen in the absence of wind transport. Wind transport also results in the sublimation of blowing snow in transit. Pomeroy and snow researcher Eric Brun from France observed that wind redistribution of snow is the primary process of desertification in northern steppe environments like those in Russia. The same desertification effect has been observed in the Arctic. Russian studies demonstrated that up to 70 per cent of the annual snowfall in parts of Siberia is eroded by blowing snow. In 1995 Pomeroy and Gray demonstrated that snow-cover erosion rates for fallow fields in the southern Canadian prairies could be as high as 77 per cent. They also demonstrated that sublimation of snow during transport increases with downwind fetch. Other studies indicated that 25 per cent of the amount of snow being carried by wind can sublimate if transported over distances of 1 kilometre. Even more astonishing is evidence that suggests that up 75 per cent of the snow being transported by wind can sublimate if it is carried over a distance of 6 kilometres.

The reason blowing snow has such a high rate of sublimation is that snow suspended in the wind has a much higher ratio of surface area to mass than snow that has not been separated from the pack. It has been found that, depending on wind speed, air temperature, relative humidity and to a lesser degree solar radiation, a unit mass of blowing snow can have as much as 3000 times more area exposed to the atmosphere than snow that remains within the snowpack on the ground. Temperature, again, is a critical value in the sublimation equation. It has been found that sublimation rates can increase

25-fold when the air temperature above a snowpack rises from −35°C to 0°C, which happens often and very quickly in places like the chinook belt that spreads across the southeastern slopes of the Rocky Mountains and extends outward to parts of the prairies.

Sublimation takes a great deal of water out of the snowpack. At −15°C and at a relative humidity of 70 per cent, the sublimation loss from snow particles suspended 10 metres above the snowpack in wind blowing at 20 metres per second, or roughly 70 kilometres an hour, will be the snow-water equivalent of 1 millimetre of rainfall an hour. At such rates of moisture loss it does not take long to greatly reduce or even eliminate the water reservoir effect of a given snowpack.

Pomeroy and Gray calculated annual moisture losses to wind transport at four locations in Saskatchewan. They found that on average at least 8 to 11 per cent of annual snowfall is removed by wind from a 1-kilometre-long fetch of 25-centimetre-tall crop stubble. On the southern prairies the percentage of annual snowfall loss to wind transport nearly doubled, to 19 per cent. They discovered that snow transport nearly doubled again in fields left fallow, resulting in an additional 7 per cent loss of moisture as a result of sublimation. They calculated that the average percentage of annual snowfall loss to sublimation at the four Saskatchewan sites ranged from 23 to 41 per cent of annual snowfall. Additional information from a total of 16 different sites on the Canadian prairies revealed that the average annual amount of snow water lost to sublimation from a 1-kilometre fetch was up to five times the amount of snow transported to the field edge.

Though Pomeroy and Gray did not comment on the implications of this finding for the agricultural future of the Great Plains, those are clear and worrisome. Extensive open fields of agri-monocultures with standing stubble harvested for biofuels increase the potential for blowing-snow sublimation resulting in reduced water availability and soil moisture. Our agricultural field practices, in combination with the draining of some 70 per cent of all wetlands that once existed on the Canadian prairies, may be making the entire region drier over time. If precipitation increases that are projected to occur as a result of warming prairie temperatures do not compensate for increased evaporation and sublimation, climate change may exacerbate this drying. Though it is always tricky to predict the future, it is not impossible that feedbacks among these factors will lead to accelerated desertification of parts of Canada's southern prairies.

In the late 1990s Pomeroy worked with colleague Phil Marsh to examine sublimation rates farther north in Canada. They found that blowing-snow erosion tended to decrease as one advanced north and transitioned from grassland to boreal forest. In the High Arctic, however, they found that the same wide-open, unobstructed spaces that generated a lot of snow transport on the prairies also resulted in a great deal of snow sublimation in tundra regions. In 1997 Pomeroy and Marsh estimated that 27 per cent of the annual snowfall sublimated into vapour as a result of wind transport in the transition zone between subarctic and arctic in the Mackenzie River basin in the Northwest Territories.

In tundra regions north of the treeline the figure was even higher. Nearly a third of all the moisture that fell in the tundra of the western Arctic was lost to sublimation that took place during the transport of snow by wind. This loss contributes significantly to the fact that much of the High Arctic is in fact a cold desert. Marsh, whom we will meet again in chapter four about advancing IP3 research in the cold regions of the North, would spend most of his career studying wind-related snow transport, snowmelt and related cold regions hydrology parameterization problems in the western Arctic. In the meantime, John Pomeroy would go on to undertake further research into wind-borne snow transport. He would also do new research on the role that the interception of snowfall by vegetation plays in snow ecology and hydrology.

## MODELLING SNOW THAT DOESN'T REACH THE GROUND

Snow interception in the mountainous regions of western Canada is largely controlled by accumulation of falling snow in the canopy of coniferous forests. Like other accumulations of snow, this snow is subsequently affected by sublimation, melt, drip from melting snow and the unloading of snow by the branches of trees that form the canopy. Snow intercepted by the forest canopy can therefore arrive on the ground below as a solid, a liquid or a vapour. As John Pomeroy has proven, the importance of these specific processes in governing snow cover varies with local weather patterns, regional climatic circumstances, tree species and the density of the forest canopy. He has also found that the amount of sublimation that takes place in the canopy has a huge influence on forest hydrology, an influence the average person might find astounding.

The modelling of these hydrological processes requires a clear understanding of the parameters that define how much snow can accumulate in the forest canopy as opposed to on the ground. To know how snow accumulation would affect the hydrology of an entire forest region, Pomeroy had to first understand how snow accumulated on a single tree. This meant understanding what happened at the level of a single representative branch. Pomeroy and Gray and collaborator R.A. Schmidt of the US Forest Service determined that there are three principal parameters that affect how much snow will accumulate at the branch level in a given canopy.

The first is the extent of elastic rebound of snow crystals falling onto individual branches and onto snow already held by the branch. The greater the rebound, the more likely a snow crystal would fall off the branch and land on lower branches or on the ground. Schmidt found that rebound from branches happens most often near the edge of the branch, which means that large branches lose proportionately less snow to rebound than small branches do. In this process too, cold matters. Elastic rebound is most pronounced below −3°C and declines rapidly as the temperature rises from −3°C to the melting point, at 0°C.

The second parameter that defines the amount of snow captured in the forest canopy is the extent to which the branches themselves will bend under a load of snow. This matters because bending decreases the horizontal area of the branch and increases the angle of its downward slope, which increases the likelihood that falling snow crystals will rebound from the surface. The greater the elasticity of the branch, the greater the bending that will occur as the snow load increases. The colder it is, the more ice crystals will form on a given branch. Ice crystals act as a cold scaffold, reducing movement. Pomeroy found that branch elasticity increases linearly with temperature.

The third defining parameter in determining how much snow will be captured by a forest canopy is the strength of the snow structure itself, which again is a function of temperature and humidity. The degree to which snow holds together on a branch is directly related to the degree of bonding, or strength of the interlocking snow crystals. Snow structures are generally stronger when under colder temperatures. As temperatures rise, snow tends to metamorphose into simpler structural arrangements with fewer bonds between snow crystals.

These three parameters together show that greater interception of snow takes place in the forest canopy with lower temperatures, because under such conditions snow has a lower density, crystals adhere better to one another and cold-hardened branches bend less under the accumulating weight of captured snowfall. This becomes very important to the hydrology of a given forest or forested region because the vertical distribution of intercepted snow in the lower 10 to 20 metres of the atmosphere results in a much larger volume of air exposed to the snow. The surface area of intercepted snow exposed to the atmosphere can be between 60 and 1800 times greater than that of snow that has fallen to the ground. This greater surface area results in sometimes extraordinary rates of turbulent transfer of atmospheric heat to the intercepted snow and the rapid removal of the resulting water vapour through sublimation. This sublimation can suck up a tremendous amount of heat energy from a forest environment in the course of a winter. The large surface area and the extended period of exposure of snow-covered forest canopies during long northern winters result in sometimes extraordinary rates of moisture loss through sublimation, even in cold temperatures.

Pomeroy first discovered this astounding fact when he and his student Newell Hedstrom were conducting early research into snow capture in the forests of northern Saskatchewan. They hung a black spruce tree from a weigh scale dangling from a boom on a tower in the boreal forest (see photo on page 151). Measurements conducted on this single black spruce in Prince Albert National Park in December of 1992 and January of 1993 revealed that one-third of the snow that accumulated on this single tree sublimated when the rest eventually fell to the ground. They later found, using landscape-based snow-survey techniques, that 31 per cent of annual snowfall sublimated from black spruce and jack pine canopies in northern Canada. Further extrapolations demonstrated that about one-third of annual snowfall is lost to sublimation in dense coniferous forests throughout Canada. Further extrapolation of these estimates to the boreal forest of western Canada revealed that an average 46 millimetres of snow-water equivalent is lost to sublimation each year in a region that receives from 100 to 200 millimetres in water equivalent from annual snowfall. It would not be until after he and his colleagues completed further research at Marmot Creek in Alberta, however, that Pomeroy would fully appreciate how important the sublimation effect was in the context of upland watershed hydrology.

The extent to which intercepted snow sublimates determines the amount of snow that accumulates on the ground, which in turn directly affects how much water is available to terrestrial and aquatic ecosystems. While sublimation losses vary depending on the nature and extent of the forest canopy, the importance of these losses should not be underestimated, especially in the context of deforestation. The sublimation-causing vertical redistribution of snowfall is the mechanism through which the structure of the forest influences water supply through the medium of snow. Both snow depth and water equivalent vary in relation to the distance from trees, the distance between stands of different tree species and in relation to total winter leaf or needle area.

Thus we see that forests affect hydrology in at least three significant ways. They capture, store and redistribute water over the landscape: into aquatic ecosystems, the soil and subsurface aquifers. During periods of high physiological activity in spring and summer, forests transpire large volumes of water into the atmosphere, which becomes available as precipitation elsewhere. Through the engine of sublimation, forests also put a great deal of water back into the atmosphere, even during the coldest periods of winter. Forests are the living link between the earth and its atmosphere. They are the corals that thrive in our ocean of air.

## REFLECTING ON SNOW: THE ALBEDO OF CLEAR-CUTS

If the presence of forests is an important factor in the movement of water through the hydrological cycle, then deforestation must also influence the way water moves through the world. John Pomeroy has given a great deal of thought to how various logging practices change local and regional hydrology by reducing the amount of snowfall captured by the forest canopy.

Predictably the results demonstrate that, though the volumes will vary according to the size and exposure of the clearcut, larger amounts of snow accumulate in clearings. Research in southern British Columbia has demonstrated that the increase in accumulation can range between 4 per cent and 118 per cent compared to adjacent coniferous forests. In southern Alberta, snow accumulation in clear-cuts was seen to increase between 20 per cent and 45 per cent. It has also been found that snowpacks in clear-cuts can melt up to five times faster than in adjacent forested areas, which has been shown to result in a snow season that is two weeks shorter on cleared lands compared with surrounding natural forests. The runoff occurs faster in open areas, and less water is stored by the forest, which can lead to decreased soil moisture during prolonged periods without precipitation.

It was also found, however, that the increase in snow accumulation in clear-cut areas diminished once the open area reaches a certain size. Beyond a radius of 12 times the height of the tree, clearings retain less snow than adjacent coniferous forests. This appears to be caused by the fact that the likelihood of wind transport and the resulting erosion of the snowpack increase with the area of the open space. The combination of all of these factors is what makes the modelling of hydrological change in forested regions challenging. While some of the parameters that are required for modelling are straightforward measurements, others are not. Weather conditions, including air temperature, humidity, wind speed, precipitation and incoming shortwave and longwave radiation, can be easily measured. Parameters related to the snow cover itself, such as surface temperature and relative surface roughness can also be determined easily with today's instruments. Parameters like the reflectivity of the snow's surface, however, are much more complicated.

Because it is composed of crystals in varying states of decomposition, snow is a complex surface. During the course of a day, the light falling on snow will be of a very broad spectrum. Put the two together and what you have is a constantly changing surface responding to constantly changing light. It is as if snow and light dance with one another. What is being revealed now is that the heat that is given off in the dance between snow and light is what we have taken for granted as climate.

During this interaction, snow and light change each other. Albedo – the reflectivity of snow – is one of the most amazing of all the hydroclimatic parameters. Snow reflects light differentially over the range of wavelengths from 200 to 2800 nanometers. From 200 to 800 nanometres, the level of reflection is very high and relatively independent of the grain size of the snow. Between 800 and 1900 nanometres the level of reflection drops sharply but is very sensitive to the grain size of the snow. The larger and more rounded the grain, the lower the reflectivity of the snow. Above 1900 nanometres, the level of reflection off the snow is very low. This matters a great deal because this presents variations of albedo from 90 per cent of the light that falls on the snow to 50 per cent, which has huge consequences for the plants and animals which compose the ecosystems that exist beneath, in and above the snowpack.

Such life depends upon the presence of liquid water, which

can be produced daily in a snowpack as a result of radiation penetration. Depending on the state of the snow, radiation can enter as deep as 20 centimetres. It is this penetration that makes life – and the chemical reactions necessary to life – possible. While the low heat conductivity of snow is what insulates life in and beneath the snow from the cold, it is radiation that makes liquid water and water vapour available inside the snowpack, particularly late in the winter. This water diffuses from warmer parts of the snowpack to colder parts. It is the daily availability of this water that makes it possible for species such as snow algae to bloom in such profusion within certain spring snowpacks.

Researchers have observed that in field conditions albedo is largely a function of grain size, the extent of contamination of the snow and the roughness of its surface. Surface reflectivity is highest in fresh snow and decreases in lockstep with how much thaw has occurred within the snowpack. Refreezing of the snowpack does not appear to noticeably reduce albedo. Contaminants, however, can reduce albedo dramatically, dark-coloured materials in particular. A thin layer of volcanic ash covering just one-tenth of the area of the snowpack, for example, will decrease peak albedo from 95 per cent to 40 per cent for snow grains of 1 millimetre in diameter. While albedo in polar climates can increase over time as blowing snow shatters snow crystals, instantly reducing their grain size, albedo in subpolar climates tends to decay with the time elapsed since the last snowfall. Rates of decay are very different for shallow snowpacks than for deep ones. When snow cover is less than 30 centimetres deep, the heat of the underlying ground begins to influence reflectivity. Researchers have calculated that the albedo of snow in a given snowpack on the prairies remains at 80 to 90 per cent when it is cold and then drops to half that level when the snowpack is wet and actively melting. What this means is that a snowpack may only be absorbing 10 to 20 per cent of the sun's heat when it is cold, but it is absorbing 40 to 60 per cent of available radiation when melting. What Pomeroy found interesting is that snow intercepted in coniferous canopies has no effect on the albedo of forests, despite the high degree of reflection that may occur from individual intercepted snow clumps. The reason for this is that these vertically stacked canopies act as light traps that warm quickly when incoming solar radiation increases.

As anyone who has walked amid late spring snow in a coniferous forest will have noticed, albedo under forest canopy is also strongly influenced by the accumulation of leaf or needle litter on the snow surface and by reduction of the area of snow cover during melt. Other factors affecting albedo include the extent of cloud cover and the density of the forest canopy itself. The amount of shortwave radiation reaching the ground under a dense canopy can be an order of magnitude less than that striking the top of the canopy. It is for this reason that snow can linger so long in dense coniferous forests. The persistence of snow cover in thick forests provides a form of water storage which can prolong streamflows long into the spring and even early summer, which again influences hydrology, not just locally but over entire watersheds.

## HOW MUCH WATER IS ON DEPOSIT IN THAT SNOWBANK?

The question many scientists and thinking politicians are asking is how the lower snowfall accumulations that have been projected in many climate-change scenarios will affect the supply and security of Canada's water. It appears that lower accumulations in many areas will have noticeable effects on hydrology regionally. Hydrological modellers have identified two main factors that may be affected by decreased seasonal snow cover in Canada: the insulating effect snow has on the ground below; and the cascading effects that decreasing snowfall will have on ecosystems that sustain soil, on aquatic ecosystems and on the larger ecosystems that rely on meltwater for their vitality.

Within dry snowpacks, temperatures are usually stratified in a manner that reflects the condition of the snow that has accumulated as a result of different snowfall events. Because the warmth released from the soil is greater than that provided by the atmosphere above, warmer temperatures tend to lie beneath colder ones from the time the snowpack forms until late winter and spring when the atmospheric temperature warms. In midwinter the difference between the temperature of the atmosphere above and that of the soil beneath can be as much as 50°C in shallow subarctic and arctic snowpacks. It has been demonstrated in the French Alps that a snowpack 50 centimetres deep will keep the bottom snow layer near 0°C from mid-November to mid-December. Reduction in the depth of the snowpack, however, reduces the insulation provided to the bottom layers and the soil. In a 20-centimetre snowpack the bottom layer freezes to −8°C. This effect would be compounded by the increase in the number of rain-on-snow events that is projected to accompany the reduction of both the depth and the extent of snow cover in most climate-change scenarios for the northern hemisphere. Midwinter rainfalls increase the density and reduce the depth of established snowpacks. Astounding as it may seem, a doubling of the density of a given snowpack in such circumstances reduces its thermal insulation by a factor of eight!

A smaller, shallower and more intermittent snowpack will also act differently during the spring melt. In late winter and spring, alpine, subarctic and arctic snowpacks can undergo melt in top layers while the lower layers remain frozen. In such snowpacks liquid water produced by surface melt will permeate the snowpack and be subject to daily refreezing. The presence of liquid water reduces the viscosity of the wet snow, resulting in further compaction of the snowpack. This compaction is further accelerated by the metamorphosis of plate-like snow crystals into crystals that are spherical in shape. If wet snow conditions prevail for more than a week, the density of the snow can increase dramatically. The density of a wet snowpack can be 350 kilograms per cubic metre after a night of freezing but increase to 550 kilograms per cubic metre later in the same day if the snowpack has been primed with meltwater.

At night, refreezing is usually confined to the top 20 centimetres of the snowpack. Initially, refrozen ice bodies within the snowpack are formed of large crystalline melt–freeze ice honeycombs up to 5 millimetres in diameter. As the freezing continues, these structures become hard, rigid and impermeable. The density of these ice structures within such snowpacks can result in weight concentrations as heavy as 800 kilograms per cubic metre. While making the snow easy to walk on, densities of this order also alter the weight-bearing capacity of the snowpack itself, which, in mountainous terrain, often results in springtime wet-snow avalanches that strike with the force of falling concrete. In snow-dominated mountainous areas, the paths of annual avalanches define habitat as clearly as if they had been cut with a knife.

Warming-related shifts in ecosystem dynamics caused by hydrological change will not just affect mountainous regions. In combination with an increased number of rain-on-snow events, shallower snowpacks covering smaller areas will accelerate freeze–thaw cycles in all regions where snow cover currently persists. Over time this will affect existing vegetation regimes in ways we cannot yet anticipate, especially in places where snow cover becomes more transient. Depending on how quickly these changes take place, some species will be able to adapt. Where individual species are already at the limits of their tolerance or range, however, a greater number of rain-on-snow events and a resulting shallower snowpack will invite the northward advance of species better adapted to warmer conditions. A rapidly shallowing snowpack covering a smaller area for less of each year will also constitute the kind of ecological disturbance that will invite the introduction of a greater number of invasive microbial, insect, plant and animal species.

As John Pomeroy and Eric Brun noted in their chapter on the physical properties of snow in *Snow Ecology*, the snowpack not only affects ecology but is also affected by it. "No other feedback between physical and biological systems," they wrote, "is so strong, evident and persistent around the world." The validity of this statement was confirmed again ten years later when the IP3 network released stunning new findings from a research project undertaken between 2004 and 2010 at Marmot Basin under the aegis of the Canadian Foundation for Climate and Atmospheric Sciences. These findings were immediately recognized as being an important development in the history of the Kananaskis area of the Rocky Mountains.

## STUDYING SNOW AT MARMOT CREEK

As Gillean Daffern has pointed out in the 4th edition of her classic *Kananaskis Country Trail Guide*, scientific research has played a major role in defining what the front ranges of the Rocky Mountains in southern Alberta mean to those who live there and have been visiting there for nearly 80 years. The valley became a centre for science in 1934 when the Dominion of Canada's Forest Experiment Station was opened on Highway 40, which branched off the Calgary–Banff Highway to access the Kananaskis valley. The site was used as a relief camp in the 1930s. Because of its relative remoteness, the Forest Experiment Station was later chosen as the location for a prisoner-of-war camp during the Second World War. German POWs and Canadian conscientious objectors waited out the war in a stunningly beautiful setting, building forestry roads and trails and working on the Barrier Reservoir. Following the war the bunkhouses from the camp were moved to locations on the newly completed "Wonder Road" between Lake Louise and Jasper, where they served as youth hostels. Scientific research resumed in Kananaskis in 1952 and continues to this day. In 1966 the University of Calgary's Kananaskis Field Station was built next to the Forest Experiment Station. It is now known internationally as the home of the U of C's Biogeoscience Institute. The Coldwater Centre is now housed in the former offices of the Forest Experiment Station.

While the Biogeoscience Institute is engaged in research throughout the Kananaskis region, the work conducted by John Pomeroy and his colleagues as part of the IP3 initiative was concentrated in a small watershed adjacent to the Nakiska ski area some 20 kilometres south of the offices and laboratories located at the Institute. The site has an interesting history.

Under the auspices of what was called the Marmot Basin Project, researchers at the Forest Experiment Station initiated a hydrological study of the Marmot Creek basin in 1962. This study was part of one of the largest watershed studies in Canada for over 25 years, an operation led by the Canadian Forestry Service. The larger experiment that was conducted in the Marmot Creek basin involved the logging of the forest in various places using different techniques to see what the effects might be on snowpack depth, snowmelt patterns and resulting streamflow. In the 1960s Marmot Creek became part of the International Hydrological Decade and had substantial involvement from the Alberta government and federal agencies such as the Inland Waters Directorate, which later formed part of Environment Canada.

The results of the early research undertaken in the Marmot Creek basin were much in the news in the early 1980s during preparations for the Winter Olympic Games held in Calgary in 1988. The province of Alberta wanted to create a ski area outside the national parks that could be used as an Olympic venue. A number of sites were examined, including Tent Ridge, Fortress Mountain and an area adjacent to Marmot Creek. Marmot Creek, however, was the only site to have received exhaustive scientific attention with respect to measurement of snowfall, snowpack and seasonal snow cover. The findings of that research suggested that the lower Kananaskis valley was a poor location for a ski area because of unreliable snowfall and the frequent appearance of warm winter wind systems called chinooks that sometimes sublimated and melted the snowpack right to ground level within hours. In the end, however, the decision on the location of what would become an Olympic site was political. The science was ignored, and while the Olympic events went off with few hitches, the Nakiska ski area has had to rely for the last 20 years upon expensive snow-making technology to provide a reliable base for skiing (see map on page 152).

Fifty years after snow research began in the Kananaskis, the current Marmot Creek research site is still located next to Nakiska. Research now being conducted there is part of a broader collaboration between the University of Calgary's Biogeoscience Institute and the Centre for Hydrology at the University of Saskatchewan in Saskatoon, where Dr. John Pomeroy holds a Canada Research Chair in water resources and climate change.

In the latest iteration of research conducted under the

auspices of IP3, the processes being examined by Pomeroy and his team at Marmot Creek include, as we have seen, the dynamics of snow accumulation, the effects of wind on the transport of snow, the interception of snow in forest canopies, and the effects of sunlight on snowmelt under vegetation. The parameterizations this project aimed to improve included those related to the accurate characterization of the influence of blowing snow over complex mountain terrain, snowmelt under forested subcanopies, snow interception losses, the impact of chinook winds on sublimation and melt, and the delineation of areas that contribute to runoff during snow melt. The object of these investigations and efforts at improved parameterization was improved prediction of how wind affects hydrological processes through snow transport; overall improvement in the level of complexity that models can represent; and the creation of new parameters that can be incorporated into existing hydrological models.

## WHAT HAPPENS WHEN SNOW MELTS?

Pomeroy and his colleagues have made some very interesting discoveries about how runoff is generated from snow, delivering meltwater to the base of the snowpack. Runoff in the Rocky Mountains appears to be due, more than anyone expected, to differences in internal snowpack energetics and snow hydrology that occur as a function of depth of the snowpack (see graph on page 153).

Pomeroy's research has demonstrated that snow isn't a surface in the strictest sense of the term. Rather, it is a porous medium. He has discovered that a great deal of sublimation can occur in blowing snow. As noted earlier, sublimation losses can average as much as 30 per cent in blowing snow in the alpine regions of the Canadian Rocky Mountains. But Pomeroy found that sublimation losses of 50 to 60 per cent occur in mountain forest snowpacks in the Kananaskis area as a result of frequent chinooks sublimating snow directly from that intercepted in the forest canopy. This is one of the highest winter sublimation rates in the world. The discovery of this fact alone demonstrates how vulnerable winter snow can be to rising temperatures. We have no idea as yet of how a continuing rise in mean annual temperatures will affect the frequency and intensity of chinooks in the front ranges of the Rockies. We do not, therefore, understand what changing winter weather patterns will do to winter snowpack and how this will ultimately affect water supply.

The presence of forests also has a great influence on snow interception and subsequent sublimation and unloading to the ground from branches. Dr. Pomeroy has been able to illustrate how longwave radiation from the sun is transferred from heated foliage to the subcanopy. He and his team are now able to accurately parameterize and predictively model these processes. In the next step in this team's work, researchers want to better characterize parameters related to snow unloading from forest canopies and the hydrological implications of wind transferring snow from one basin to another and accumulating it in drifts. Each researcher involved in the IP3 initiative has been working to make simulation more consistent with observation so that our water-resource and climate models are accurate and reliable. Pomeroy wants to make these parameterizations available for practical use in the broadest range of modelling situations (see photo on page 154).

## THE IMPORTANCE OF MARMOT LAB

IP3 research basins are also teaching venues in which professors supervise the work of graduate students seeking advanced degrees in hydrology and related hydroclimatic specializations. In research conducted by the University of Saskatchewan's

Chad Ellis, under the supervision of John Pomeroy in collaboration with IP3 partners Richard Essery of the University of Edinburgh and Tim Link of the University of Idaho, Ellis found that radiation is the main energy source for snowpack warming and melt in coniferous forests in mountain regions. Utilizing extensive field observations, Ellis and his colleagues quantified the effect of needleleaf forest cover on radiation and snowmelt timing in pine and spruce forest sites and nearby clearings of varying slope and aspect in Marmot Creek. They found that, in pine and spruce environments alike, forest cover acted to substantially reduce total radiation to snow and thus delay snowmelt timing on south-facing slopes, while increasing total radiation and thus advancing snowmelt timing on north-facing slopes. Their results strongly suggest that impacts of radiation on snow and snowmelt timing from changes in mountain forest cover depend a great deal on the slope and aspect at which these changes occur (see photo on page 155).

When the IP3 initiative came to a close, Pomeroy and his colleagues assessed their findings in the context of contemporary understanding of what is happening to the snow-dominated hydrology of the Rocky Mountains in Canada. The first thing they discovered was that their enhanced parameterizations made it possible for their simulations to better reproduce what actually happens in nature. They found that an energy-balance snowmelt model coupled to a physically based blowing-snow model on a cold regions hydrological model platform is capable of reproducing observed alpine snow accumulation, redistribution and ablation regimes. The results of this modelling prove, once again, that cold matters.

The modelled sensitivity of the mid-alpine ridge snow regime to winter warming during the study period suggests warming of up to 4°C will reduce blowing snow transport dramatically and reduce maximum snow accumulation by up to two-thirds. As a result, snowmelt runoff may be reduced by almost half. Temperature rise of up to 4°C will also increase actual evapotranspiration by up to 10 per cent and will more than double the rainfall runoff as a result of an increase in rain-on-snow and rain-on-bare-ground events. The spring runoff season will move forward by one month and the amount of water discharged by the basin will be reduced by about 10 per cent. As Pomeroy put it, a cold regions hydrology basin is about to become more temperate.

In addition, IP3 researchers at Marmot Creek discovered that winter warming in the Rockies is already associated with reduced streamflow, which suggests that we may expect less water to be produced by spring and summer snowmelt in western Canada in the future. The implications of this finding for water security in Alberta, Saskatchewan and Manitoba are profound.

Professor Pomeroy explained that a central finding of the IP3 research project was that Canada's hydrological systems, at least in the country's cold regions, are on the move. The natural variabilities of rain- and snowfall, river flows, lake levels and the extent of the country in which permafrost persists are all changing. In scientific terms, Canada is experiencing a loss of hydrological stability. As a result, precipitation and river flows will be different than we have come to expect. New ranges of variability will emerge. There will be more and more times when that variability will be outside the range for which our urban and rural infrastructure has been designed to function. There will be more times when climate variability will be outside our current ability to adapt. As Pomeroy and his colleagues discovered, these changes are already underway in the mountain West. They are even more pronounced, however, in the Canadian North.

# CHAPTER 4

## *The Impermanence of Permafrost: A North No Longer Frozen*

Dr. Andrew Weaver is one of Canada's most respected climate scientists. Weaver is a professor and Canada Research Chair in climate modelling and analysis in the School of Earth & Ocean Sciences at the University of Victoria. He was also a lead contributing author to the UN Intergovernmental Panel on Climate Change Fourth Assessment Report, published in 2007, and was among the elite international group of UN-linked scientists who shared the Nobel Peace Prize with climate activist Al Gore that same year.

In his 2008 book *Keeping Our Cool: Canada in a Warming World,* Weaver provides convincing evidence to suggest that since 1979, natural forces have in fact acted to cool our planet, while human activities have acted with greater influence to warm it. He then demonstrates how scientists have been able to formally detect the effects of human-caused warming against the backdrop of natural climate variability. Weaver argues it is incontestable that human combustion of fossil fuels has led to increased warming of the lower atmosphere and the upper ocean; sea level rise; an increase in the frequency of heat waves and the intensity of tropical cyclones; surface air pressure changes; and increased specific humidity. Weaver also observes that while there has been an increase in total precipitation in some regions, this has been accompanied by a greater likelihood of drought and forest fires in other places. While Weaver notes most of the changes in terms of increases, he also notes one really large and obvious decrease: the extent of Arctic sea ice. The reduced extent and thickness of sea ice may be a parameter that may be feeding all of the increases that are causing concern over climate change.

The effects that are causing the Arctic to warm faster than lower latitudes are well known. As snow and ice melt, darker land, lake and ocean surfaces absorb more solar energy. More of the extra trapped energy goes directly into warming than into evaporation. Because the atmosphere is thinner in the Arctic, it takes less heat to warm the air enough so that it warms the surface. As sea ice retreats, solar heat absorbed by the oceans is more readily transferred to the atmosphere. The warming of the oceans can alter atmospheric and oceanic circulation in ways that further accelerate warming. It is a feedback loop that no one wants to see get out of control.

In his 2010 book *A World Without Ice*, climatologist Henry Pollack explains that ice plays a major role in setting the temperature of the earth's atmosphere and oceans. In so doing, it governs major weather patterns and regulates sea level. In this regard cold matters a great deal in that there is not a person on earth that is not affected somehow by the moderating influence of ice on global mean temperatures.

Pollack worries that climate change impacts that lead to accelerated exploitation of the Arctic could be the destructive act that pushes the global climate system out of its tenuous equilibrium and results in runaway warming. To understand how this could happen, he explains, it is important to appreciate the differences between the Arctic and Antarctic ice caps, despite the apparent symmetry of ice at the high latitudes of the earth's poles. As Pollack points out, the two regions are indeed "poles apart" in terms of ice dynamics. The south pole is located about 3000 metres above sea level and 1300 kilometres inland on a huge continent at the bottom of the world. The north pole, on the other hand, is about 4500 metres below sea level and more than 800 kilometres from the nearest coastline. The north pole lies beneath a 4- to 6-metre-deep sheet of frozen ocean, while the south pole lies under 3000 metres of glacier ice. The glacier ice at the south pole moves at perhaps 10 to 12 metres a year. The ice at the north pole, however, can move with ocean currents at a speed of 5 to 10 kilometres a day. Because of the nature of its ice and its near proximity to adjacent continents, the north pole is more accessible to development. Antarctica virtually doubles in area when the Southern Ocean freezes each austral winter. Because of its remoteness, its difficult access by sea and the depth of ice that covers it, whatever resources may exist on the Antarctic continent are not likely to be exploited anytime soon.

The key point Pollack makes, however, is about the importance of water in the determination of climate. Water in one form or another, he explains, is the central element in our planet's thermostat. The earth's thermostat does not, however, make daily adjustments to temperature as a thermostat in a house might. Instead, adjustments take place over millions of years and are related to temperature-sensitive geological processes. The earth adjusts its thermostat downward by means of geological processes such as weathering, which results in dissolved chemicals being carried to the sea where they combine with carbon to produce calcium carbonate, or limestone, sediments of which accumulate on the ocean floor. When less weathering takes place – such as when the continents are covered with glaciers in an ice age – the carbon builds up in the ocean, which means oceans cannot absorb as much carbon dioxide from the atmosphere, causing concentrations in the atmosphere to increase and temperatures to rise.

As Dr. Pollack explains, the earth's temperature "oscillates between warmer and cooler through fluctuations in the effectiveness of the natural atmospheric greenhouse, which in turn modulates the availability of calcium to form limestone. But the functioning of this thermostat is dependent on oceans full of water in which to absorb $CO_2$ from the atmosphere and deposit limestone on the ocean floor. Without water, earth's thermostat would be broken."

## PLANETARY-SCALE WATER TRANSFERS AS AN "ORDERLY MARKET"

Orderly transfers of water, Pollack explains, are important to climate stability. In financial markets, Pollack observes, transfers of capital between one market segment and another can be accommodated without upsetting the market unduly, "provided the transfers are in small enough parcels spread out over reasonable periods of time." Orderly financial market transactions follow this kind of pattern. But if huge blocks of stock in one sector are suddenly dumped onto the market all at once, the market can be overwhelmed, producing undesirable instability.

The market analogy, Pollack explains, is useful for thinking about the transfers between ice and ocean reservoirs that are so central to the functioning of the earth's hydrological system. As glacier ice melts in response to a slowly changing climate, meltwater forms streams that merge into rivers which eventually reach the sea. The fresh water of the rivers is gradually mixed into the saltwater of the oceans and the great hydrological account book will show the balance in the ice account slowly going down and the balance in the ocean account slowly increasing. In these situations we see an orderly shift of hydrological capital from one reservoir to another.

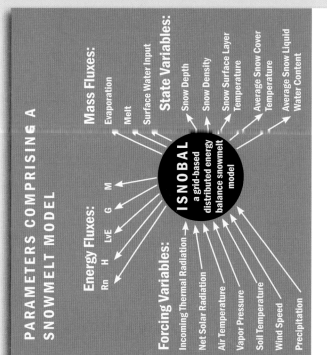

## PARAMETERS COMPRISING A SNOWMELT MODEL

**Energy Fluxes:**
Rn  H  LvE  G  M

**Forcing Variables:**
Incoming Thermal Radiation
Net Solar Radiation
Air Temperature
Vapor Pressure
Soil Temperature
Wind Speed
Precipitation

**ISNOBAL**
a grid-based distributed energy balance snowmelt model

**Mass Fluxes:**
Evaporation
Melt
Surface Water Input

**State Variables:**
Snow Depth
Snow Density
Snow Surface Layer Temperature
Average Snow Cover Temperature
Average Snow Liquid Water Content

*Like other snow process models, ISNOBAL requires a precise understanding of a broad array of parameters (see page 43).*

Courtesy of the Western Canadian Cryospheric Network

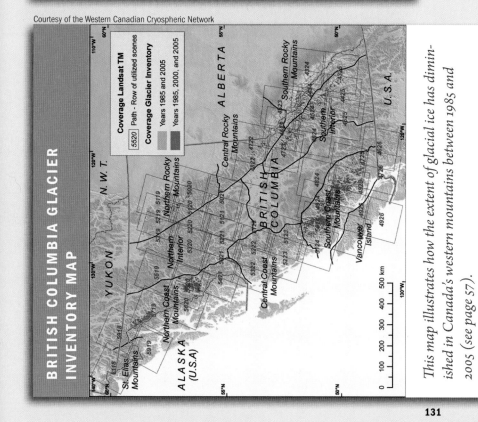

## BRITISH COLUMBIA GLACIER INVENTORY MAP

**Coverage Landsat TM**
5520 Path - Row of utilized scenes

**Coverage Glacier Inventory**
Years 1985 and 2005
Years 1985, 2000, and 2005

*This map illustrates how the extent of glacial ice has diminished in Canada's western mountains between 1985 and 2005 (see page 57).*

131

*Geological Survey of Canada glaciologist Mike Demuth measures the snow accumulation at Pole 140 on the Wapta Icefield just above Peyto Glacier in Banff National Park, Alberta (see page 59).*

Photograph courtesy of R.W. Sandford,
UN Water for Life Decade, Canada

*Instruments and a sleeping hut above the Peyto Glacier research site on the Wapta Icefield in Banff National Park (see pages 58–60).*

Photograph courtesy of R.W. Sandford, UN Water for Life Decade, Canada

*The left photograph of Saskatchewan Glacier, an outflow of the Columbia Icefield, was taken by Austin Post in 1964. The photo on the right, of the same glacier, was taken by the author in 2009. What is important about these images is that they demonstrate that glaciers in Canada's mountain West are clearly receding. What is not as obvious is that they are also down-wasting at a rapid rate as well. In other words, their length and volume are diminishing simultaneously.*

Courtesy of Austin Post and Bob Sandford

*This photograph of Snowbird Glacier in Banff National Park illustrates how a single glacier can break apart during periods of prolonged melt to produce two glaciers. (In 2009 the tail of Snowbird separated from its body.) In periods of persistent warming, the number of glaciers can actually increase briefly, but each of the separate new glaciers is smaller and more susceptible to complete disappearance.*

Photograph by R.W. Sandford, UN Water for Life Decade, Canada

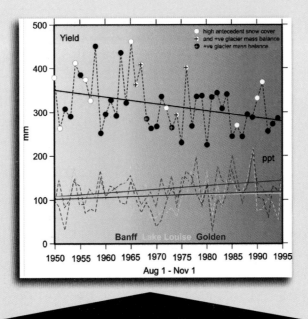

Research has shown that glacial masses (top graph, "Yield") have been declining for the important period of late summer to early winter despite the fact that precipitation (bottom graph, "ppt") was on a modest rise for the region during those same months for the period of record.

IP3 glacier research was conducted adjunctly at the Geological Survey of Canada's site at Peyto Glacier in Banff National Park. GSC/Natural Resources Canada and Environment Canada have been doing mass balance, glacier hydrology and climate change research at this site for some 50 years, following on from work done here earlier in the 20th century by the organizations that eventually became NHRI Glaciology and GSC Glaciology.

Image courtesy of Dr. Alain Pietroniro, Water Survey of Canada, and Dr. Mike Demuth, Geological Survey of Canada

Changes in the extent of glaciers in the Mount Robson area to 2005. This map is based on what is known about the extent of ice at Robson at the close of the Little Ice Age, a cool period between about 1550 and 1850 CE when glaciers last advanced in the Rocky Mountains. The other dates represent the years for which information was extrapolated from later topographical maps, aerial photographs, trim line studies or satellite images to complete the map.

Image supplied by Dr. Roger Wheate, WC²N

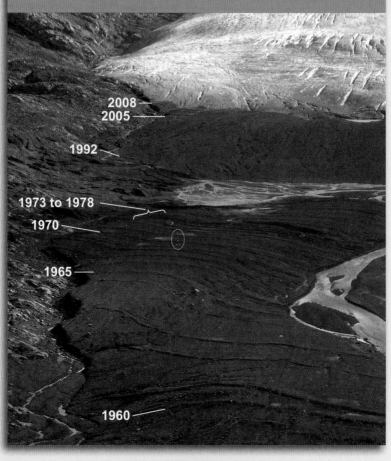

## PUSH MORAINE STUDIES ON CASTLE CREEK GLACIER.

2008
2005

1992

1973 to 1978

1970

1965

1960

*Moraines are composed of rocky debris deposited by a moving glacier. A record of the historic ebb and flow of a given glacier can be derived through careful examination of the patterns in which its moraines were deposited. Castle Creek Glacier in British Columbia is clearly retreating at a rapid rate.*

Photograph published in *Geophysical Research Letters,* October 2009.
Image courtesy of Matt Beedle, WC²N

## THE RAPIDLY DIMINISHING COLUMBIA ICEFIELD, 1920–2000: ELEVATION CHANGE (m)

*Changes in extent of icefield, 1920 to 2000:*
*yellow = 1920; red = 2000; black = water divide.*

One of the more interesting discoveries made through the Western Canadian Cryospheric Network relates to a more accurate assessment of the exact area of the Columbia Icefield, which straddles the Great Divide between Alberta and British Columbia at the boundary of Banff and Jasper national parks. While the assessment does not fully account for how much ice may be buried beneath moraines, researchers now estimate the area of the Columbia Icefield to have been about 223 square kilometres in 2005, considerably smaller than the 325 square kilometres alleged in tourism literature and local lore. What is also interesting is the extent to which the Columbia Icefield has diminished in depth as well as surface area (see page 66).

Courtesy of Christina Tennant, WC²N

The Columbia is one of nine major glaciers that flow from the Columbia Icefield. Large masses of ice such as the Columbia Icefield are known to have a cooling effect on regional climate. In midsummer the difference in temperature between the ice and the surrounding rock walls can be as much as 30°C. As glaciers shrink, the landscape warms faster and stays warm longer, which affects patterns of snow cover and depth of snowpack. Without the refrigerating effect of glaciers, more precipitation is likely to fall as rain in winter, further affecting streamflows in mountain rivers. The Columbia Glacier is the headwaters of the Athabasca River, which flows through Alberta's oil sands.

R.W. Sandford, UN Water for Life Decade, Canada

*Above is a 1920 photograph of Columbia Glacier and the upper Columbia Icefield taken in Jasper National Park by a member of the Alberta–British Columbia Boundary Survey Commission.*

*Below is Columbia Glacier as it looked in September 2009. Notice its much-diminished length and thickness and the huge lake that now exists at its snout as a result of meltwater being dammed by the 1920 moraine.*

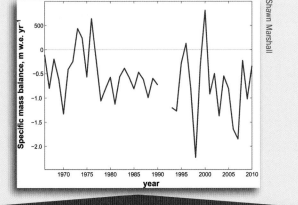

*Peyto Glacier in Banff National Park is one of the most-studied bodies of ice in Canada. This graph indicates a general trend of decline in mass over the past 40 years (see page 70).*

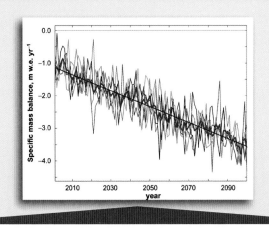

*This graph projects the rate of decline of Peyto Glacier that can be expected to occur if the climate trends of the past 40 years persist until the end of this century.*

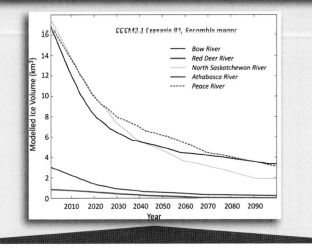

*Dr. Shawn Marshall and his colleagues projected streamflow declines in five major western rivers over the coming century (see page 70).*

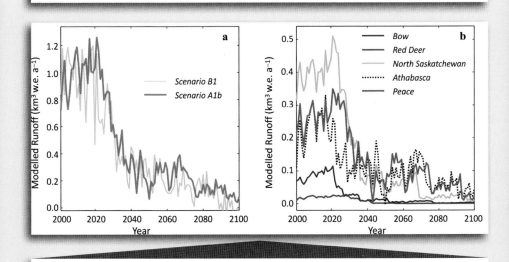

*Projected runoff in the Bow, Red Deer, North Saskatchewan, Athabasca and Peace river systems under conservative climate change scenarios.*

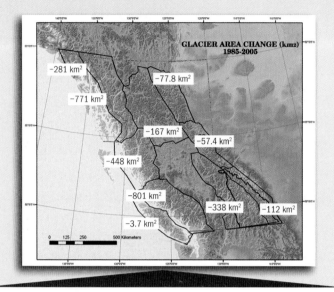

GLACIER AREA CHANGE (km2)
1985-2005

−281 km²

−77.8 km²

−771 km²

−167 km²

−57.4 km²

−448 km²

−801 km²

−338 km²

−112 km²

−3.7 km²

0   125   250        500 Kilometers

*Above: Glacier area change in British Columbia.*
*Below: Glacier area change in British Columbia and*
*Alberta as percentage of glaciated area (see page 68).*

Courtesy of the Western Canadian Cryospheric Network. This work was published in the journal
*Remote Sensing of Environment* (January 2010).

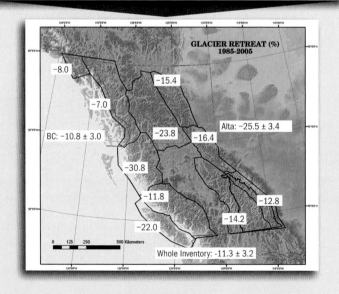

GLACIER RETREAT (%)
1985-2005

−8.0

−15.4

−7.0

Alta: −25.5 ± 3.4

BC: −10.8 ± 3.0

−23.8

−16.4

−30.8

−11.8

−12.8

−14.2

−22.0

0   125   250        500 Kilometers

Whole Inventory: -11.3 ± 3.2

*What the St. Elias Mountains in the Yukon look like now, and what they might look like in the absence of the massive icefields and huge glaciers that exist today.*

Courtesy of Dr. Garry Clarke, WC²N

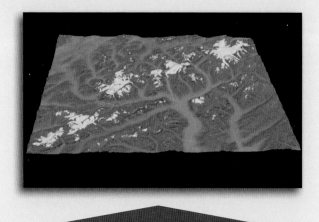

*WC²N estimation of the icefields and glaciers in the area of the Icefields Parkway in Banff and Jasper national parks as of 2002 (see page 88).*

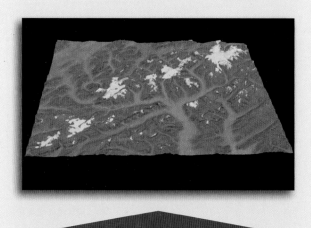

*2040 estimation.*

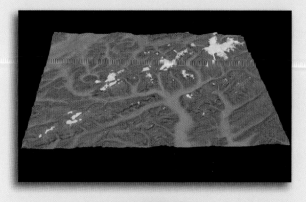

*2080 estimation.*

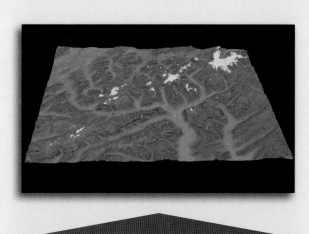

*WC²N estimation of the icefields and glaciers that may still be in existence as of 2100, under a variety of climate-change scenarios.*

Courtesy of Dr. Garry Clarke et al., WC²N

Employing a specialized suite of mathematical tools, Garry Clarke and his colleagues were able to roughly characterize the physical nature of the landscape beneath Athabasca Glacier and to put the ice back on top of the resulting illustration as at 2002 in order to project how long the glacier might survive under various climate warming scenarios.

2040 projection.

2080 projection.

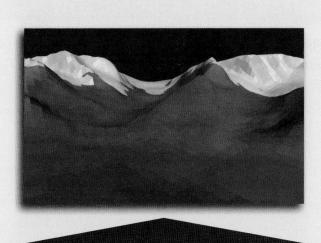

*If the projections put forward by Clarke and his colleagues prove correct, not much will remain of Athabasca Glacier by 2100.*

Courtesy of Dr. Garry Clarke et al., WC²N

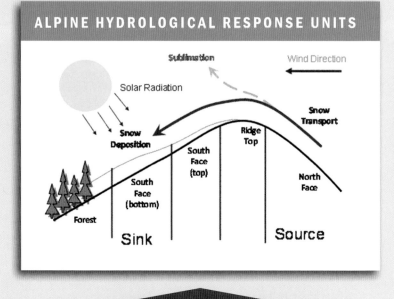

## ALPINE HYDROLOGICAL RESPONSE UNITS

Sublimation

Wind Direction

Solar Radiation

Snow Transport

Snow Deposition

Ridge Top

South Face (top)

South Face (bottom)

North Face

Forest

**Sink**

**Source**

*The direction of prevailing winds has a significant impact on patterns of both snowfall and snow redistribution. This diagram illustrates the variety of hydrological response units modellers have to consider in examining the hydrology of mountain ecosystems (see page 105).*

Courtesy of Dr. John Pomeroy, IP3

*One of the ingenious methods John Pomeroy and his colleagues at Marmot Creek devised to determine the effects of snow accumulation at the canopy level was "the hanging tree." They suspended a tree of a known weight from a tower in the forest and weighed the accumulation of snow before, during and after each storm.*

Photograph courtesy of Dr. John Pomeroy, IP3

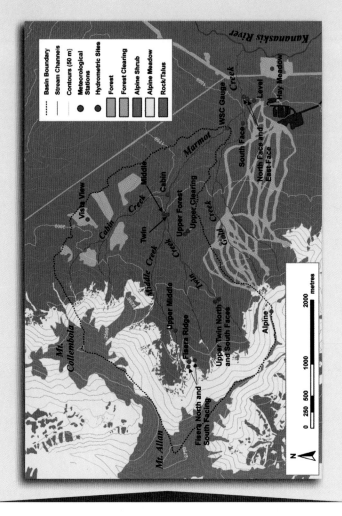

The Marmot Creek drainage was chosen as a research site because it possessed all the biogeographic elements of a typical mountain headwaters. The watershed has its origins in the bare rock and snows of the upper alpine. The creek then flows through the alpine into Hudsonian forest before descending into the montane zone in the lower Kananaskis valley (see page 122).

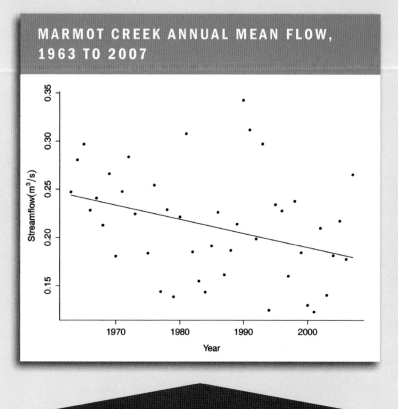

## MARMOT CREEK ANNUAL MEAN FLOW, 1963 TO 2007

Researchers Phillip Harder and John Pomeroy plotted the streamflow of Marmot Creek from 1963 to 2007. They demonstrated that average spring and summer flow declined by approximately 13 per cent in that 44-year period. It appears that changes in precipitation, snowpack depth and the extent and timing of snow cover are already reducing the water supply that originates in the Rocky Mountains (see page 123).

Image courtesy of Phillip Harder and John Pomeroy, IP3

Hydrological research in mountain terrain requires considerable physical stamina. Field technician May Guan and research assistant Logan Fang use a snow density measurement tube and a specially calibrated scale to record the weight of the snow at Fisera Ridge, high above Nakiska ski hill in the Marmot Creek research basin (see page 124).

Dr. John Pomeroy downloads image data from a time-lapse camera at the IP3 Fisera Ridge site in the Marmot Creek research basin in Kananaskis Country (see page 125).

Photograph courtesy of Lynn Martel

The IP3 research site at Peyto Glacier. The Peyto has been studied by the Geological Survey of Canada for more than 40 years. It is one of the world's reference glaciers and an important surrogate in hydrological studies in watersheds fed by glaciers throughout western Canada. The instrument and research hut are located on a moraine above the glacier (see page 173).

Photograph courtesy of R.W. Sandford, UN Water for Life Decade, Canada

# A GAUGE NETWORK IN DECLINE

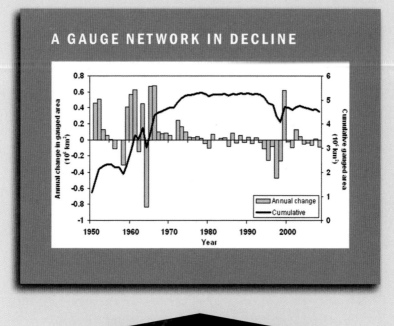

Careful and continuous monitoring of hydrological processes is central to providing the basic data which science needs so that it can offer informed perspectives on the state and fate of our water resources. Since the 1990s, monitoring programs have been eliminated as a result of budget cuts in almost all federal and provincial jurisdictions. Accurate field monitoring is the foundation of good forecasting and prediction. Without reliable field measurements, it will take longer to perfect the ecohydrological models we need to develop if we wish to know how best to adapt to climate impacts on our country's water supplies (see page 185).

Image courtesy of Dr. Stephen Déry, International Polar Year

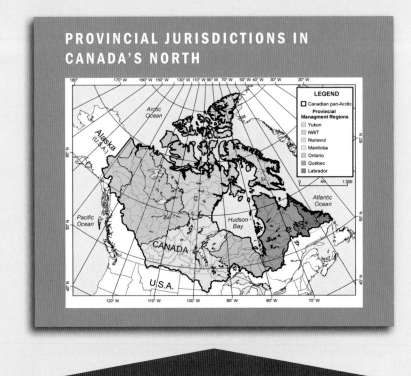

There is a great deal of southern interest in the Canadian North. British Columbia, Alberta, Saskatchewan, Manitoba, Ontario, Quebec and Newfoundland/Labrador also have jurisdiction over vast northern areas. It has been projected that hydrological monitoring of these northern reaches would have to be increased by at least two orders of magnitude just to achieve the very minimum monitoring standards recommended by the World Meteorological Organization (see page 186).

Image provided by Dr. Stephen Déry, International Polar Year

# GROWING ANTHROPOGENIC INFLUENCES

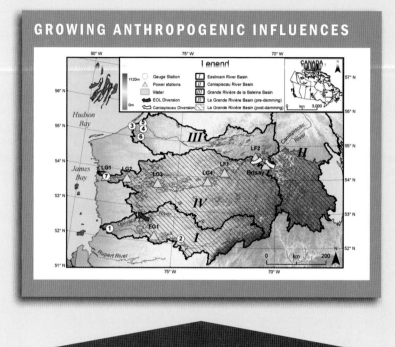

The extent of human influence on northern regions is already considerable and will expand dramatically as the resources that lie beneath the snow and ice become more easily available. Hydro power development is also likely to be substantial as we see in this illustration of dams, power stations and major water diversions in the James Bay area of northern Quebec. Sustainable northern development will require that the ecohydrological effects of such developments be well understood so that mitigation measures can work and human adaptation to such changes is possible (see page 186).

Image provided by Dr. Stephen Déry, International Polar Year

| Stationary climate and coping range (variation around a well-defined average) | Climate change (variation around a trend) | Vulnerable |
| Adaptation |
| Coping range |
| Planning horizon | Vulnerable |

*The fixed envelope of relative certainty within which we have come to expect our weather and climate to fluctuate over a typical period of time. The range of possibility depicted is what engineers, for example, would consider in designing major infrastructure such as bridges or highways. Also shown is the increasing number of extremes we will have to cope with over time. This loss of what is called "stationarity" is in fact making our society increasingly vulnerable to accelerating climate change effects. Ian Burton, Environment Canada's lead expert on climate change adaptation, coined the term "adaptation deficit" to characterize this failure to recognize or fully acknowledge the evidence that our climate system and its related services are moving in a direction that is beyond the range of our current capacity to adapt.*

Mike Demuth

*The upper reaches of Haig Glacier in late summer. All of the previous winter's snow has melted and the glacier is being wasted away by melt (see page 72).*

Photograph by
Dr. Shawn Marshall

*Ice roads such as this one are very important in NWT. Heavy loads can be transported to remote mines and communities can receive supplies throughout the winter without expensive air transport. With warming, ice roads cannot be constructed until later in the season and sometimes need to be sprayed with water to ensure they can bear the weight of the vehicles that use them. Some worry that ice roads will not be reliable in the future if warming continues (see page 210).*

Photographs courtesy of R.W. Sandford, UN Water for Life Decade, Canada.

*Great Slave Lake in* NWT *at spring breakup. The Mackenzie River basin is increasingly seen as an ecohydrological linchpin holding the ice–water–weather–climate interface together in the northern hemisphere.*

Photograph by R.W. Sandford,
UN Water for Life Decade, Canada

There are times, however, when abrupt shifts in the movement of hydrological capital cause chaos in the exchange. One such moment of turmoil occurred during the melting of the last continental ice sheet in North America when meltwaters that were flowing by way of the Ohio and Mississippi rivers suddenly found a new and more direct way to the sea, creating what is now the St. Lawrence River. Large volumes of cold meltwater no longer flowed into the Gulf of Mexico via the Mississippi but instead coursed northeastward past the Gaspé Peninsula into the Atlantic Ocean. There, these cold new inflows interjected themselves into the northward drift of the Gulf Stream, interrupting the delivery of heat to the North Atlantic. A chill fell over Europe for a thousand years.

The Younger Dryas, as this period was called, lasted until the melt of the North American ice sheets was reduced and the volume of cold flows became insufficient to maintain a thermal barrier to the Gulf Stream, which then resumed the delivery of heat to Greenland, Iceland and Western Europe. The fear at present is that the disappearance of Arctic ice in combination with accelerated melt of the Greenland ice sheet may cause a similar disruption to the climate of the northern hemisphere. Too huge a block of fresh water could suddenly be dumped onto the hydrological market all at once and climate turmoil could ensue. If the Arctic sea ice disappears it will be the first time in 55 million years and thus will be the largest change in the earth's surface ever experienced by human beings. Because our knowledge of the earth's climate system is still comparatively limited, we can't predict what the impacts may be.

As Henry Pollack points out, water in each of its iterations affects or has the potential to affect climate. Permafrost covers 20 per cent of the earth's surface and is in some places two kilometres deep. But in regions where average annual temperature is only slightly below freezing, the permafrost is much thinner. The loss of this permafrost will result in further release of methane into the atmosphere, potentially exacerbating climate-change effects.

## THE IMPORTANCE OF SEA ICE

Dr. Pollack adds that ice also forms a sediment beneath the seafloor, in a crystal structure that is in the shape of a not quite spherical cage. The depth at which this form of ice occurs is relatively shallow, within the sedimentary deposits. This particular type of ice is widespread on the continental shelves and on rare occasions in deep lakes on the continents, such as in Siberia's Lake Baikal. What makes this ocean floor ice of special interest is its ability to trap methane – natural gas – inside the molecular cage. Drilling into the shelf sediments has retrieved samples of this gas-bearing ice from a great many sites around the world. When ignited, Pollack says, the samples put on quite a show – to see chunks of ice aflame defies all intuition. Methane, of course, is an important source of energy in the industrialized world, but it is also a potent greenhouse gas when found in the atmosphere. The release of methane from its icy subsurface cage is therefore a worry in the context of runaway climate change and further sea level rise.

Pollack observes that the extent and thickness of Arctic sea ice are both diminishing at ever-faster rates. Less Arctic ice in summer means more ocean water is exposed, which in turn means more solar radiation will be absorbed, resulting in ice forming later and later in the fall. The newly formed sea ice will be thinner when it melts the following spring. Spring breakup will come earlier, resulting in a longer warming season for ocean water. This warming eventually mixes into the deeper ocean, which leads to sea level rise through thermal expansion. This warming could also influence the circulation of currents in the Arctic Ocean, with a cascading effect on hydroclimatic processes we didn't know existed or have yet to parameterize.

It all comes down to the fact that cold matters to how much solar radiation our planet can reflect. The Arctic Ocean is actually the smallest of the world's oceans, and changes in its albedo – the amount of light reflected from its surface – directly determine its temperature, which in turn defines the habitats and the lives of the creatures found there. The Arctic ice pack has thinned by 40 per cent in the last half-century and is still receding. At present, Arctic ice acts as a mirror, reflecting 95 per cent of the sun's energy. But without it, 90 per cent of that energy will instead be absorbed. Since temperature variations between the equator and the poles help shape ocean currents and jet steams alike, melting at the poles threatens massive disruption of the entire global weather pattern.

Other observers, such as Richard Ellis, author of *On Thin Ice: The Changing World of the Polar Bear*, have observed that the loss of sea ice in the Arctic Ocean is the biggest and most rapid change in the conditions of Arctic ice humans have ever witnessed. While comparable climate shifts have happened before, Ellis notes, they occurred over tens of centuries, not tens of years.

The idea that the fabled Northwest Passage would soon become free of ice as a result of global warming was first put forward as a prediction by climate scientist Jim Hansen in 1981. Solar variability, which some scientists and many climate skeptics claimed was the cause of present-day warming, did not counteract greenhouse warming as was hoped. Undeterred by attempts at government censorship, Hansen and a number of colleagues published a paper in 2007 called "Climate Change and Trace Gases," in which they made the following claim:

Paleoclimate data show that the earth's climate is remarkably sensitive to global forcings. Positive feedbacks predominate. This allows the entire planet to be whipsawed between climate states... Inertia of ice sheet and ocean provides only moderate delay to ice sheet disintegration and a burst of added global warming. Recent greenhouse gas emissions place the earth perilously close to dramatic climate change that could run out of our control, with great dangers for humans and other creatures.

Since 1979 the equivalent area of two Quebecs of sea ice has been lost, which may be what is creating the feedback loop that is accelerating further calamitous ice losses. Canadian journalist Ed Struzik was in the Arctic during the summer of 2007 when the so called "mortuary" of old ice that normally makes McClintock Channel in the High Arctic impassable disappeared. In his 2009 book *The Big Thaw: Travels in the Melting North*, Struzik quotes scientists who conjecture that 2007 may have been the tipping point for ice loss over most of the Canadian Arctic. He points out that the temperature differential between the poles and the equator is the genesis of cold fronts that bring snow and rain to more southerly climes. If polar ice shrinks in area, winters will not go away but will become warmer, making drought and extreme weather events more common in the south. Andrew Weaver, Henry Pollack, Richard Ellis, Ed Struzik and even Jim Hansen have something in common. The information they have relied upon to create and validate their perspectives has come to light in part through research projects such as those funded by the Canadian Foundation for Climate and Atmospheric Sciences, which supported the work of scientists in the IP3 and WC$^2$N networks.

## CLIMATE CHANGE AND PERMAFROST: A NORTH COMING UNGLUED

For bioclimatic purposes, the upper elevations of the mountain ranges of southern Canada can be considered a southward extension of the harsh environment of the Arctic. Disappearing ice and snow in places like the Alps and the Rockies is already telling us that we are having significant impact on the global atmospheric system, which in turn will affect other processes of landform erosion and change.

The Canadian Foundation for Climate and Atmospheric Sciences and the International Polar Year fund of the Department of Indian & Northern Affairs supported six IP3 research basins in the Canadian North. These included Wolf Creek in the Yukon; Havipak, Trail Valley, Baker and Scotty creeks, all in the Northwest Territories; and Polar Bear Pass in Nunavut.

At an IP3 conference held in Whitehorse in the fall of 2009, researchers explained that a great many mountains that were once covered in snow and ice are already darkening and drying. Such mountains reflect less light and retain more heat. Often, as a consequence, the ice that is holding them together disappears. The result is greater instability. We are already observing this trend elsewhere in many of the planet's high mountain ranges.

We already know that even before global warming has finished reducing the length and depth of our glaciers it will already be after our mountain snowpacks, with huge potential impact on everyone who lives downstream. Changes in snowpack, however, are just the beginning of changes that will converge on us over time. One of the many uncertainties relates to permafrost.

We know very little about the role permafrost plays in the larger geomorphological context in the relatively extensively studied Alps, let alone in the Rockies, the Andes or the Himalayas. We are not talking here about environmental changes that we know enough about to accurately predict. This is a concern for researchers like Dr. Bill Quinton.

Quinton holds a Canada Research Chair at the Cold Regions Research Centre of Wilfrid Laurier University. His work is focused on the hydrology of cold regions: the high-latitude and high-altitude regions where the presence of snow and ice above and beneath the ground are of profound importance with regard to the cycling and storage of water and energy. Dr. Quinton has observed that substantial knowledge gaps in cold regions hydrological processes severely limit our ability to accurately predict variations in flows and storages needed to properly manage Canada's freshwater resource. Quinton argues that this is a serious deficiency given that Canada is the steward of approximately 20 per cent of the world's fresh water. Annual revenue from hydroelectric power alone is $15-billion, which is about one-third of the net economic contribution to Canada. He is concerned that this resource is at best sparsely measured throughout northern Canada.

An immediate gain from this research is the development of new methods to quantify the freshwater resources in Canada's cold regions. This research continues to lead to the development of new process algorithms that form the basis of numerical models explicitly designed for application in polar regions. These models substantially improve the accuracy of hydrological prediction, which helps in flood planning; sound mine, road and oil pipeline design; water resource allocation; hydroelectric power station operation; and assurance of reliable sources of drinking water. As well, the models have an interface to operational weather prediction, climate and eco-systems models, so their future applications can directly contribute to estimating impacts of a warming climate and human disturbance on water availability and quality in Canada's North.

Because of the work of Bill Quinton and others in the IP3 research group, however, we do know what is happening in the North and in the Arctic. Researchers have discovered that much of the permafrost at the southern boundary of its extent in the North has a mean annual temperature of −0.1 to −0.2 °C, which means that very small temperature changes will have a huge effect on the physical landscapes and biogeography of the North. The permafrost in what are called peat plateaus is thawing, causing the land to subside over vast areas. This is already having a considerable impact on roads, bridges and airport runways and potentially on pipelines. In parts of the Arctic, buildings are beginning to tilt and sink into ground that is no longer solid because the permafrost that once made it so is melting. Cold matters. The fact that permafrost is on the way out in the North is not just a headache for those who would build pipelines down the Mackenzie valley. It means not only that the positive feedback of generating more greenhouse gases has begun but also that the hydrology of the entire Arctic is on the move.

Not surprisingly, there is a great deal of permafrost, not just in the Arctic but in the mountain regions of southern Canada. We don't know where it all is, however, and we are not studying it. In the absence of our own research, what is happening in the Arctic should provide us with valuable clues about what might happen here.

We can expect landforms to begin to change rapidly as permafrost melts, soils become more saturated and more extreme weather events result in increasingly intense precipitation events. From the work of Panya Lipovsky and the Yukon Government we can begin to appreciate the kinds of landform changes we can expect in the future Rockies.

Dr. Bill Quinton of Wilfrid Laurier University and his colleagues undertook to determine rates of permafrost loss in the IP3 research basin in Scotty Creek in the Lower Liard River valley, NWT, near the Liard's confluence with the Mackenzie. In this region, permafrost exists only beneath black spruce forest canopy; open areas such as bogs and fens are permafrost-free. Using the black spruce coverage on the aerial and satellite images as a surrogate for permafrost, Quinton and his team estimated that permafrost covered 70 per cent of the study area in 1947; 55 per cent in 1970; 53 per cent in 1977; 49 per cent in 2000; and 43 per cent in 2008. At the rate of thaw that occurred between 1947 and 2000, Quinton and his team estimated that permafrost would persist in the study area for 90 years. However, at the accelerated thaw rate measured between 2000 and 2008, the data suggests permafrost at the study site may be gone by 2040, or in only about 30 years.

The next question Dr. Quinton wanted to answer was whether or not basin runoff would respond to permafrost thaw, a question he thought might be answered through the imaging of electrical resistance of the active layers above the permafrost and in the permafrost itself.

Future work will focus on answering this question by identifying the key factors controlling the rates and patterns of preferential thaw leading to permafrost degradation. Quinton also intends to develop a new model that will simulate permafrost response to climate warming and human disturbances. He then hopes to couple the hydrological and permafrost models so as to be better able to predict the future distribution of permafrost and the effects of permafrost reduction on river flow regimes under a number of climate-change and human-impact scenarios.

In the Yukon, infrastructure, property and water quality have been significantly impacted by permafrost-related landslides in the past. There they have discovered that permafrost-related landslides can be far-reaching, can remain active for decades, can occur on gentle slopes of all aspects and can impact large areas, up to 200 hectares. They have also discovered in the Yukon that large-scale landscape change can be triggered by human and natural causes including road construction, river erosion, forest fires, heavy rainfall and confined groundwater flow, all of which are things that occur in the Rockies and other western mountain ranges. Researchers have now made it clear that basin characteristics and surficial material stratigraphy several kilometres upslope of any area targeted for development should be evaluated for permafrost landslide risk. It has also been determined that development on alluvial fans should carefully consider the risk of debris flows.

No one now doubts what is happening. Design for stream crossings in the Yukon now incorporate the potential for massive debris torrents. Still, there is a great deal that remains to be known. Maps and models of permafrost distribution and hydrological parameters (rainfall intensity thresholds for critical levels/rates of infiltration, saturation and runoff) still don't exist for most of the Yukon, even though it has been well recognized that they would greatly benefit landslide susceptibility modelling and risk management. The landscapes of Canada's cold regions are simply too big and there aren't enough resources to do more.

While it would be easy to surmise that because these lands are so sparsely populated, landform slumping in northern regions won't be a problem. Unfortunately, however, permafrost doesn't just exist in the Arctic. Because of the work of scientists like Brian Luckman, Mike Demuth and WC²N's Shawn Marshall, we have a foundation for understanding what is happening in our mountains, but even these scientists would be the first to say that we don't know enough. What they do say, however, is that what is happening in the Arctic is also happening at altitude in Canada's southern mountain ranges.

## THE GLACIOLOGY OF CLIMATE CHANGE: THE ARCTIC–ALPINE LINK

Unless you have travelled to both places, you may not notice that there is very little difference between the conditions of the Arctic and those of the high altitudes in Canada's western mountains. For reasons of altitude, the Rockies and all the other high ranges in Alberta and British Columbia can, for ecological and glaciological purposes at least, be seen as a southward extension of the conditions that exist much nearer the pole. Because of the high cost of travelling in the remote regions of the Arctic, glacial research is often undertaken in the south and the findings corroborated with much more limited glacial research conducted in the High Arctic. While patterns of precipitation may be different than in the Arctic, the relative accessibility of glaciers in the mountain national parks has made them the object of some of the longest research projects in the history of glaciology. The biggest and best-known glacial mass in Canada is the Columbia Icefield. It, along with other, smaller icefields and many hundreds of glaciers similar to those found in the Arctic, can be reached by car in only a few hours from airports in Edmonton or Calgary.

With the exception of major work undertaken there by Dr. Brian Luckman of the University of Western Ontario, however, most of the evidence we have relating to changes in glacial dynamics in the Canadian Rockies does not come from the more easily accessible Columbia Icefield area as one might expect. It comes instead from research conducted over the past 40 years at the IP3 research site at Peyto Glacier, located in the upper reaches of the North Saskatchewan basin in Banff National Park. While Peyto Glacier currently covers about 12 square kilometres, it has lost 70 per cent of its volume in the last 100 years. Winter snow depth and the duration of cover have been declining since the 1970s (see photo on page 156).

The Canadian Rockies have experienced a 1.5°C increase in mean annual temperatures over the last 100 years. During this time, increases in winter temperatures have been more than twice as large as the temperature increases in spring and summer. The annual temperature record in the Canadian Rockies is dominated by winter conditions, which show the largest inter-annual range, 12.7°C, and the greatest warming trend, 3.4°C, over the past century. In spring and summer both of these figures are more modest: the range of variation is only 4 to 5°C, the warming trend only 1.3°C.

Proxy records from tree ring chronologies recovered by Brian Luckman from a site near Athabasca Glacier at the Columbia Icefield indicate that summer temperatures were below late 20th century values from about 1100 until 1800 CE, and that the coldest summer conditions occurred during the Little Ice Age advance in the 19th century, which averaged 1.05°C below the 1961–1990 mean. Over the past century, the area of glacial cover in the Rockies has decreased by at least 25 per cent and glaciers have receded in length to approximately where they likely were some 3000 years ago, before the Little Ice Age advance. Late 20th and early 21st century recession is seen to be exceptional in the history of Rocky Mountain glaciers over that 3000-year period, which happens to be the period for which we have the most complete record.

Research at Peyto Glacier has revealed that this area of the Rockies experienced a below normal snowpack between 1922 and 1945. A period of above normal snowpack followed for the 30 years between 1946 and 1976. Comparisons of these observations with other western snowpack records revealed similar snowfall patterns in the interior and coast ranges of British Columbia, which suggests that much of western North America is responding to climate change in a coherent way.

Dr. Brian Luckman of the University of Western Ontario has been engaged in paleoclimatic research in the Canadian Rockies and other mountain ranges around the world for more than 40 years. One of Canada's most respected earth scientists, Luckman has shared his work with two generations of students and other researchers, including many of those involved in both the IP3 and WC²N initiatives.

Photograph by R.W. Sandford, UN Water for Life Decade, Canada

There was a sharp decrease in snowpack throughout the Pacific Northwest beginning in the 1970s, which proved to be consistent with inter-decadal variation in sea surface temperatures in the Pacific Ocean, called the Pacific Decadal

Oscillation. This suggests that snowpack in the broader Pacific Northwest region is influenced by ocean current movement associated with the El Niño Southern Oscillation. This oscillation expresses itself most distinctively through the presence of a semi-permanent annual low-pressure pattern called the Aleutian Low.

Evidence suggests that it is the strength of the annual appearance of the Aleutian Low that determines the extent of snowfall as far inland as the main ranges of the Rocky Mountains. In other words, what happens in the Pacific very much affects the Rockies.

Now that we know that the Aleutian Low, the Pacific Decadal Oscillation and Rockies snowfall are linked, it is possible to reconstruct earlier climatic regimes as a means of extending the record upon which predictive modelling can be developed. Recent research suggests that both maximum and minimum mean annual temperatures have been rising and that there has been less snow at lower altitudes in the Canadian Rockies over the last 20 years. Because temperature records are almost all from elevations between 640 and 1340 metres, it is suspected that extrapolations are probably underestimating warming at high altitudes. We simply don't know how much faster warming is occurring at higher altitudes, because we have no baseline and no data, which again suggests the need for more expanded hydrometeorological monitoring in the high Rockies. We do know, however, that the main ranges of the Canadian Rockies experienced higher average maximum temperatures between 1920 and 1940 and again in the 1960s, '70s and '80s. In the 1980s temperatures began rising, to reach extreme levels in 2003, the same year the Alps experienced severe warming. Years like 2003 are very dangerous. We know what extreme temperatures did to the Alps, which in places lost 10 per cent of glacial mass in only one summer. We are only now beginning to understand the impacts here.

Extreme maximum temperatures in the Canadian Rockies increased by 4 to 5°C in the early 20th century. A century later, in 2003, extreme minimum temperature increases as reported from recording stations were in the range of 7 to 8°C. No wonder mountain guides crossing the Wapta Icefield reported that for the first time in living memory the snow on the icefield itself was melting even in the middle of the night.

Troubling trends continue to emerge even after winters during which we have received close to what is considered historically normal snow accumulations. Recent very hot summers are troubling because greatly increased summer temperatures burn off even normal snowpacks by July, resulting in greater glacial loss and longer periods of low streamflow in our rivers and creeks. The fear, of course, is that extreme temperature events of this magnitude could push our climate system out of equilibrium and that past variability will no longer be a guide to the future. The problem, as climate scientists have already indicated, is that we are not ready for these kinds of changes.

## WHEN THE MORTAR OF MOUNTAINS MELTS AWAY

We don't need to convince people who are active in the mountains that what is happening in the Arctic is also happening at altitude in southern Canada. Anyone who has spent any time outdoors will know that big changes have occurred in the Rockies and the interior ranges in only one generation. Climbing routes in many parts of the mountain West are changing faster than guidebooks can keep up. The ice that held our mountains together is melting.

As the authors in *Darkening Peaks: Glacier Retreat, Science and Society*, edited by Ben Orlove, Ellen Wiegandt and Brian

Luckman, point out, there has been a noticeable increase of superficial rockfall from deglaciated rock walls worldwide. Rockfall has become a particularly serious problem on north faces. In some mountain regions whole walls are coming apart. On December 14, 1991, a rock avalanche on Mount Cook in New Zealand reduced the peak's altitude by some 20 metres when 14 million cubic metres of rock peeled off the summit. When the rock avalanche smashed into the glacier below it created its own earthquake of 3.9 on the Richter scale. But the mountain wasn't finished yet. The rock avalanche smashing into the glacier created a moving rock and ice mass of some 55 million cubic metres that travelled 7.4 kilometres downvalley at speeds up to 400 to 600 kilometres an hour.

These types of events are occurring elsewhere as well. On the evening of September 20, 2002, an ice and rock avalanche occurred on the northern slope of the Kazbek massif in the Russian Caucasus. Beginning as a slope failure below the summit of the massif, the slide slammed into Kolka Glacier, almost entirely entraining it. The rock and ice of the avalanche and the ice that composed the glacier created a debris flow of some 100 million cubic metres that blasted down the Genaldon valley for 19 kilometres before being halted at the entrance of the Karmadon Gorge. A mudflow, however, continued downvalley for another 15 kilometres, stopping only 4 kilometres from a nearby population centre. The combined impact of the rock and ice avalanche and the mudflow was devastating. About 140 people were killed and most of the region's roads, buildings and infrastructure were destroyed. But that wasn't the end. The ice dam created at Karmadon Gorge resulted in the creation of several new lakes, some up to five million cubic metres in volume, which remain an imminent threat to downstream areas.

In Europe, climate scientists are also anticipating increased rockfall associated with the accelerating advance of rock glaciers. It appears that 40 to 60 per cent of the ice in a rock glacier can be considered frozen fossil groundwater from the early Holocene or Pleistocene. These glaciers appear to be on the move.

Accelerated rockfall is also being accompanied by increased debris flows on recently unfrozen slopes. Ice-cored moraines are becoming unstable. Just as we expect an increase in the frequency, intensity and duration of extreme weather events as a consequence of rising mean temperatures, we can also anticipate more intense and frequent debris flows after these storms. At the same time, increased glacial melt is creating more glacial lakes and enlarging existing ones. Small ice-contact lakes are also forming as a consequence of glacial recession within the physical confines of recessional moraines. Unfortunately, more of these lakes are forming behind moraines that have melting ice cores, dramatically increasing the risk of massive outburst floods. These outburst floods have become a real hazard in the Alps and the Himalayas. In the Alps people can afford to pump water out of these lakes to reduce the threat. In the Himalayas they cannot; people there are in most cases forced to simply wait for the inevitable.

What we are also beginning to see is an increase in catastrophic landscape failures as a consequence of these combined influences. This is even happening in Canada. On August 12, 1997, the lower part of Diadem Glacier in the southern Coast Mountains of British Columbia fell into Queen Bess Lake. The glacier slamming into the lake produced a train of large waves. The waves overtopped the moraine at the east end of the lake and flooded the valley of the west fork of Nostetuko River, creating a flood that devastated the valley and had impacts for a hundred kilometres downstream.

A similar landform failure occurred near Meager Creek Hot Spring, about 95 kilometres north of Pemberton, BC, in August of 2010.

## IP3 IN NWT AND YUKON: FIGURING OUT WHERE "SQUARE 1" IS

But it is not just our landscapes that are on the move. Ecosystems too are moving quickly, trying to keep up with changing hydroclimatic conditions. The IP3 research initiative also concerned itself with how fast this was happening. To do that, however, more had to be known about the baseline conditions against which change can be measured. Models had to be improved based on better parameterization of the natural processes at work in places like the Arctic, where ecological changes appeared to be occurring the most rapidly. As part of a multi-pronged approach to understanding the extent of ecological change that is occurring in the cold regions of Canada, IP3 researchers experimented with a variety of models to determine which were capable of best reproducing what they witnessed in nature.

Dr. Stephano Endrizzi is a post-doctoral research scientist with the Centre for Hydrology at the University of Saskatchewan who came from Italy to work on IP3 under the supervision of Dr. Phil Marsh. He uses a hydrological model called GEOtop to examine patterns of ground thaw and runoff generation in vegetation-covered permafrost terrains. Much of Endrizzi's research is focused on the development of algorithms that accurately simulate the movement of heat in permafrost regions so that resulting landform changes can be understood and predicted. Dr. Endrizzi focused his attention on advances in the modelling of how snow cover varies over time and space in IP3's Trail Valley Creek research basin in the Northwest Territories. Once again the purpose of the study was to characterize how snow is captured and retained and to improve the parameterization of these processes at very small scales so as to enhance the accuracy of models and thus enable better predictions of what might happen to snow conditions as the climate continues to warm in the future.

From Dr. Endrizzi's work it is clear that science is not always a lot of fun. A great deal of painstaking data collection formed the foundation of every one of the research projects funded by the Canadian Foundation for Climate and Atmospheric Sciences, and the IP3 initiative was no exception. The slow and careful collection of data is not what the average Canadian thinks about when and if they ever consider the challenges of contemporary science. The collection of such data is simply plain hard, slow work. In addition, working conditions in some of the areas in which data needs to be collected are difficult and sometimes even dangerous. Dr. Endrizzi's work also made it clear that a great deal more research will be required to reconcile what actually happens on the ground in the Arctic with what has historically been understood to happen in analyses made by way of airborne remote-sensing techniques. This is a problem not just in Canada but everywhere in the cold regions of the circumpolar North.

Stephano Endrizzi is in good company in his interests in cold regions hydrology. Dr. Stephan Pohl came from Germany to Saskatoon and earned a Ph.D. in hydrology from the University of Saskatchewan in 2004. Dr. Pohl then went to work as a post-doctoral research scientist with the National Water Research Institute in Saskatoon. Between 2006 and 2010 Pohl cooperated in what he called "Project IP3," testing what is called the MESH model in an area of continuous permafrost in the IP3 research basin at Trail Valley Creek. His findings indicated that though late-lying snowdrifts covered only 8 per cent of the Trail Valley Creek basin, they held up to 33 per cent of the end-of-season snow-water equivalent. The modelling indicated that late-lying drifts augment the usually low early summer flows in the Trail Valley Creek basin and keep the water table downstream of their location relatively high.

The results of the work completed under the aegis of IP3 indicated that the MESH model Pohl was using to understand ecohydrological processes in the Trail Valley Creek basin was able to accurately simulate spring snowmelt. Although this proved not to be the case with summer rainfall-induced runoff events, which the model tended to underpredict, the inclusion of snowdrifts improved simulation.

Future work, according to Pohl, will employ output data from the GEOtop model generated by Stephano Endrizzi and others as a means of determining how to optimally calibrate MESH models so that they will be able to work with as much spatial information as can be made available, while still retaining their computational efficiency. With the end of IP3, both Pohl and Endrizzi have returned to Europe and so the mechanisms to carry out this research are now uncertain.

Rick Janowicz is a hydrological researcher and forecaster for the Yukon government, a half-million-square-kilometre territorial jurisdiction that governs Canada's western Arctic. Janowicz utilized IP3 support to examine hydrological response trends observed in the Yukon in response to climate warming over the last three decades. What he found was that these trends were most obvious in examination of permafrost, river ice and glaciers.

Janowicz observed that between 1961 and 1990 summer temperatures in the Yukon had risen by between 2 and 6°C, depending on region. During the same period, winter temperatures rose by between 4 and 6°C. These temperature rises were accompanied by summer precipitation increases of between 5 and 10 per cent and winter precipitation changes ranging from

Phil Marsh is an adjunct professor of hydrology at the Centre for Hydrology at the University of Saskatchewan and a researcher at Environment Canada's National Hydrology Centre, also located in Saskatoon. He is an expert in the parameterization of snow and small lake processes at the tundra transition zone in Canada's western Arctic. His work and that of 14 IP3 colleagues was conducted at Trail Valley Creek in the Northwest Territories. The research focused principally on the parameterization of spatial variability in snow cover and changes in wind speed. There was a special emphasis in this work on the effect of the rapid expansion of shrubs into the tundra eco-zone, which is seen to be a clear indication of climate warming.

The advance of shrubs into the southern margins of the tundra is expected to be the largest change in vegetation in Northern Canada in the coming decades, making it an important new parameter in the determination of snow cover. This project relied heavily on detailed field observations funded by a number of earlier projects as well as by IP3 and the International Polar Year. The combined results of these studies were incorporated for testing in a number of simulation models, including the cold regions hydrological model (see, for example, page 22); a model called GEOtop (page 179); and the Canadian land surface scheme (CLASS) and MESH models, briefly described at page 13 and 29 respectively. Once again the object of this project was to make modelling predictions representative of what actually happens in nature.

−10 per cent in some areas to +20 per cent in others. In combination with the diminishment of permafrost, these temperature changes have resulted in increased winter streamflows due to greater groundwater contributions. Loss of permafrost has also resulted in peak flows decreasing in summer because melt now takes a longer route along ice-free pathways to reach stream channels than it did when the permafrost was in place, causing more melt to be absorbed into newly thawed soil to become groundwater. In basins which still have glaciers, such as the Atlin River, however, higher temperatures are generating greater summer glacier melt, which results in increased peak flows. The combination of all of these factors is having considerable impact on river ice regimes.

Freeze-up of the Yukon River at Whitehorse now occurs about 30 days later than it did in 1902. The timing of spring breakup as measured at Dawson City since 1896 has advanced six days over the course of the century. The same trend has been observed on the Porcupine River at Old Crow. One of the problems emerging from the tandem effects of these temperature and hydrological changes is that river ice jams are becoming more frequent at breakup, causing flooding. Another trend that is of concern is the increase in midwinter ice breakup on some rivers.

Janowicz pointed to the 2009 spring breakup of rivers in the Yukon as an example of the complicated interactions between more extreme temperatures, permafrost melt, changing hydrological regimes and river ice dynamics. The winter of 2008 and 2009 was colder than normal, which resulted in thicker river ice. Because of the influence of ocean currents, the winter snowpack was 150 to 175 per cent of normal. Spring, however, came early after colder than usual temperatures. When spring arrived there was rapid warming to record high temperatures. The breakup of the Yukon River, which normally takes 30 days, took only a week. Ice jams suddenly appeared in numerous places along the river, some of them in locations not historically prone to ice jams. Minor flooding occurred at Dawson City and Old Crow, but 150 kilometres downstream the town of Eagle, Alaska, experience a record flood.

Janowicz has dramatic pictures of what the river ice did in Eagle. It literally broke up and expanded up the banks of the river and pushed its way into the town, knocking buildings off their foundations and then pushing through their walls, flattening everything in its path. Then the river flooded the town.

Janowicz observes that climate change is not something theoretical where he lives. It is a reality in the Yukon. His work has led him to believe that teleconnections exist between rising temperatures, permafrost loss, peak streamflows, ocean current influences on winter precipitation and changing river ice volumes and breakup conditions. These teleconnections, or relationships between climate anomalies separated by large distances (typically thousands of kilometres), Janowicz argues, need to be more fully understood, as they are likely to cause more frequent midwinter ice breakups as temperatures continue to rise, resulting in greater frequency and intensity of springtime ice-jam flooding in many northern rivers.

Rick Janowicz is not alone in his fear of rapid hydrological change and its potential effect on human aspirations in the North. Dr. Stephen Déry and a team of researchers from the University of Northern British Columbia in Prince George undertook analyses of streamflow changes in response to rising temperatures in northern Canada during the International Polar Year in 2009. This work was integrated with that conducted by researchers under the aegis of IP3. Throughout the work, Déry decried the decline in the number of streamflow gauges presently installed in the north. He observed that there has been a 15 per cent decrease in the area of the North being monitored for hydrological purposes since the 1990s. He also noted that 12 additional dams were being projected for the Nelson River, which flows from Lake Winnipeg to Hudson Bay. Proper development of dams such as these would rely heavily on streamflow monitoring information that is no longer available (see graph on page 157).

Consistent with Rick Janowicz's observations, Déry indicated that streamflow in northern Canada was considerably above average during the International Polar Year in 2009. Part of the work of his team was to establish the nature of river discharge anomalies during this time and characterize the factors that led to these anomalies. The hope was that his team could provide an updated record of river discharge trends, which would certainly be useful to those who manage water resources in the North.

## FLYING ON INSTRUMENTS:
## CLIMATE MODELLING NEEDS CLIMATE MONITORING

In the Cariboo Mountains in British Columbia, Stephen Déry and his team also undertook research on blowing snow, which is an important element in glacier mass balance. Wind, snow conditions and topography strongly influence snow-mass redistribution. What they learned was critical. The understanding of these processes, Déry noted, requires either dense physical monitoring networks or high-resolution modelling to ensure accurate quantification. The problem was that without accurate monitoring as a foundation, accurate high-resolution models are very difficult to create.

Déry is not alone in this view. John Pomeroy, in his capacity as principal investigator for the entire IP3 initiative, and many other scientists as well have been arguing for some time for the need for enhanced upper-elevation and latitudinal monitoring of snowpack and snow cover. The basis of this argument has been that it is in these high places and high latitudes that we can expect the effects of climate change to be most immediate and most dramatic. It is in the high latitudes in particular that we should expect the most development to take place in the near future (see map on page 158).

There is, however, official resistance to increased funding for physical monitoring. Governments argue it is unnecessary to continue expensive long-term monitoring, while modellers such as Déry and others hold out the promise of accurate results generated almost solely by physics-based computer approximations that require only sporadic data collection from the field so as to compare what is actually happening in nature with the model results. In Déry's opinion, this should not be an either/or proposition. While really dense monitoring networks may not be necessary in some instances (even if they were affordable), more field monitoring is clearly necessary and valuable, as Déry went on to point out through the example of his team's own research at a number of high-elevation sites in the Cariboos. His team utilized direct physical monitoring to characterize and then model how blowing-snow events resulted in rapid changes in snow depth at Castle Creek Glacier. The team was then able to use that same data to accurately simulate an extreme blowing-snow event at Browntop Mountain in the Cariboos in 2006. Next, Déry and his researchers will test their hard-won parameterizations on other models (see page 159).

Environment Canada's Chris Spence and his researchers spent four years working to improve process understanding and parameterization methods that hopefully would lead in the future to better prediction of hydrological processes in IP3's Baker Creek research basin in the Northwest Territories. Spence's work builds on the research on boreal lakes done by Dr. Raoul Granger of Environment Canada, who for the first time developed mathematical methods to calculate evaporation losses from cold northern lakes and show how lake temperature controls the evaporation rate. Granger tested this method with detailed measurements of evaporation using meteorological towers on islands in Baker Creek and in northern Saskatchewan.

Unlike the IP3's Trail Valley Creek research basin, Baker Creek is in Canada's subarctic Canadian Shield. Spence and his team recognized that existing mathematical parameterizations do not adequately simulate the vast array of ecohydrological processes that occur in the Canadian Shield or any other hydroclimatic region of the Arctic. After identifying some of the known gaps in knowledge, Spence and his colleagues set themselves the difficult but critical task of creating parameterizations of natural ecohydrological processes that had never been characterized before.

Putting the work of Granger and Spence together, we see that warmer temperatures do indeed lead to more evaporation and thus to less water storage in lakes. This can result in a drastic decline in streamflow in the North, putting local hydroelectric facilities below capacity and changing not only canoe routes but the aquatic ecosystem health of a vast region.

Chris Spence and his colleagues demonstrate how much work has to be done just to establish a single scientific fact upon which even the most elementary predictions can be founded. The work of this project alone should give the public confidence in the scientific method and in the research outcomes of networks like the Improved Processes Parameterization & Prediction Network and the Western Canadian Cryospheric Network. These outcomes give us hope that sometime in the future these researchers or those who follow their scientific tracks will be able to make prediction of future hydrological circumstances possible. This should also give the public confidence that science is looking for the answers to the right questions. While industry, through applied research and efforts at commercialization, may be looking for answers to their own questions, the pure research undertaken in networks such as IP3 and WC²N are seeking answers to the questions that matter not just for commerce but for all of society.

In an ideal world, in which IP3 and WC²N would continue to be funded, testing would be continued until representations of model responses were accurately reflective of observed subcatchment dynamics. It appears, however, that this is not likely to happen any time soon. As a result we continue to develop in our warming cold regions without any clear picture of what the future might bring.

Though Spence never talks about it, he and researchers like him know they are in a race. As clearly exemplified by major economic development proposals such as the Mackenzie gas pipeline, Quebec's multi-billion-dollar Plan Nord, which will commit to the exploitation of northern resources in that province over the next three generations, and the federal government's evolving plans to assert sovereignty in the Arctic, a great deal of southern attention is being focused on northern affairs. If our country's history is any indication of what we can expect in the future, development will proceed faster than scientific understanding of our actual and potential impacts and how those might be mitigated.

Perhaps the realization that change is happening faster than we can comprehend it is the overriding cumulative finding of the research conducted by the Canadian Foundation for Climate & Atmospheric Research under the aegis of IP3 in the North. The biggest discovery may in fact be that we barely understand the world as it is now, let alone what it may be like in the rapidly approaching future in which warmer temperatures remind us again and again how much cold matters.

# CHAPTER 5

## Collective Breakthrough:
## How Canadian Scientists Revolutionized
## Our Understanding of Climate Change

Through IP3 and WC$^2$N we now know with certainty that we have lived through a relatively stable and clement period in the earth's recent climate history. Science has also confirmed that the climate has in fact always been changing but that this change in the past was slow enough that we were able to adapt successfully to it. Now we find that hydroclimate change is accelerating.

Changes are beginning to occur so quickly that natural systems are having trouble keeping up with what has recently become a persistently upward trend in atmospheric temperature. Whole ecosystems are beginning to unravel and many species are already at the limits of their adaptive capacity. We have discovered that not only does the rate of warming in cold regions count, but that cold itself matters. Many are concerned that, like the ecosystems that support us, we ourselves may not be able to adapt easily to the rate at which warming is altering our world.

Because it is a locus of immediate and dramatic climate impacts, and because of the impending opening of the Northwest Passage to marine navigation, a great deal of attention is presently being focused on the Arctic. We have learned that the

Arctic warms faster than lower latitudes because as snow and ice melt, darker-coloured land and ocean surfaces absorb more solar energy, and more of the extra trapped energy goes directly into warming rather than into evaporation. The Arctic also warms faster than lower latitudes because the atmospheric layer that has to warm in order to in turn warm the surface is shallower in the Arctic. The Arctic is also warming faster than anywhere else on earth because as sea ice retreats, solar heat absorbed by the oceans is more easily transferred to the atmosphere, and alterations in atmospheric and oceanic circulation can and do increase warming.

Other very important new knowledge is becoming available as a result of long-term research undertaken by a variety of agencies. All of this research indicates that the Arctic is going to be the place in Canada where Canadians may have to adapt first and most to climate change impacts. We can expect more high-calibre work to emerge from Canada's North. Hopefully that work will set a new tone for the kinds of adaptive responses we will have to make with respect to the effects of climate change in southern Canada.

## THE *ARCTIC CLIMATE IMPACT ASSESSMENT*

The bible of baseline knowledge concerning climate change in the Arctic is the *Arctic Climate Impact Assessment*, a compendium of research conducted by 300 scientists from 15 countries produced by the Arctic Council in 2005. At the time it was published, this 1042-page, two-kilogram, large-format book had essentially everything in it that we presently knew about the current state and fate of the circumpolar Arctic as it relates to climate change.* The book provides detailed analysis of Arctic climate in the past and present; indigenous per-

* While this book could be the source of lifetimes of interpretation and conjecture by hundreds if not thousands of people interested in the future of Canada's Arctic, it is not likely the average person will be able to afford this beautiful but expensive volume. Fortunately, its contents are available in four languages on the Internet at www.acia.uaf.edu or by Googling the words "arctic climate impact assessment."

spectives on climate change effects; explanations of how the models that were used to project future climate scenarios were developed and how they work; and detailed information on projected effects of warming on the atmosphere, on snow and ice and on freshwater and marine systems as well as on Arctic biodiversity. The volume also offers observations on the current and potential impacts of a warming Arctic on aboriginal hunting, herding, fishing and food gathering at the species level as well as the impacts climate change may have on human health. The book explores the projected impacts, region by region, that global warming will have on forestry, land management, agriculture and physical infrastructure in the circumpolar North. The work concludes with an exploration of ecological and cultural resilience in the context of current climate-change projections.

The authors observe that the climate of the Arctic has undergone rapid and dramatic shifts in the past, and suggest there is no reason why it could not experience similar changes in the future. Since the Industrial Revolution, however, anthropogenic greenhouse gases have added another climate driver. In the 1940s the Arctic experienced a warm period, like the rest of the planet, although it did not reach the level of warming experienced in the 1990s. The authors of the *Arctic Climate Impact Assessment* note that the United Nations Intergovernmental Panel on Climate Change has stated that most of the global warming observed over the 50 years between 1955 and 2005 is attributable to human activities associated with land-use changes and greenhouse gas emissions.

The latter chapters of the *Assessment* provide beautiful colour illustrations of what the Arctic is projected to look like over time. These illustrations present current boundaries of the extent of summer ice, permafrost and northern treeline as well as changes in these that are projected to occur by the end of the century. The change in the permafrost boundary assumes that present areas of discontinuous permafrost will be completely free of permafrost at some point in the future, though it is hard to estimate just when. The incontestable point these illustrations clearly make is that if current warming trends continue it will not be long before the Arctic becomes a very different place. The problem with these projections, however, is that they are likely inaccurate because the models they were derived from were based on projections of a slower rate of increase in global greenhouse gas emissions and on hydrological algorithms that no longer represent reality. Because higher temperatures are energizing the hydrological cycle, many of the standard parameters traditionally used in modelling are no longer accurate. As we will see below, the whole climate system appears to be speeding up, which is why Intergovernmental Panel on Climate Change predictions are often shown to have underestimated the rate of change that is taking place.

## JAMES LOVELOCK AND THE IMPORTANCE OF MONITORING

In his 2009 book *The Vanishing Face of Gaia: A Final Warning,* James Lovelock offered other observations, based on a lifetime of climate modelling experience, on why the IPCC climate models have been unable to adequately warn us of the accelerating extent of Arctic warming and sea level rise. In Lovelock's opinion, the IPCC models fail principally because inadequate monitoring results in unreliable predictive outcomes. We do not do enough field monitoring and that is why our predictive models are so inaccurate.

Lovelock claims that such models will never be of adequate quality, because good predictions do not come from physics alone, as we are discovering through our efforts to model the hydrological alterations that climate change is bringing about in ungauged basins in Canada. Reliance on physics-based models that attempt to make up for the absence of reliable field data by employing mathematical algorithms have consistently failed to accurately predict rises in mean global temperature, rates of increase in sea level and sea ice loss.

The second reason Lovelock believes the IPCC models are unreliable is that they don't account for Arctic ice loss, which in his estimation will cause as much warming as 70 per cent of the $CO_2$ present in today's atmosphere. The third reason Lovelock has no confidence in the IPCC models is based on his concern that the projections don't take feedbacks into account.

Lovelock has lost faith in IPCC models because in fact we can't adequately model what is happening now, let alone predict what might be happening to our planet's atmosphere 50 years from now. He believes that the reason the models are flawed or imperfect or incomplete and/or outdated is that they cannot possibly deal with an atmosphere that in his opinion is 99 per cent composed of the products of living things at the surface which keep changing in composition and nature in response to one another and to human impacts. This fact alone, Lovelock maintains, should be turning the life sciences and geology on their heads.

Lovelock posits that the models will be wrong so long as the wrong worldview prevents science from seeing the real dynamics that power the nature, evolution and self-regulation of our planet's atmosphere. He maintains that other models have to be developed and applied to the dynamics of the two great planetary ecosystems: algae in oceans and plants on land. Only in this way can models be improved enough to allow for accurate forecasting and prediction of climate impacts. The reason for this, Lovelock writes, is that there is a direct link between temperature and carbon dioxide concentration in the atmosphere and plant and algae growth. Ocean physics, he explains, determines that warm surface water separates from and floats atop the cooler waters below at temperatures greater than about 12°c and so denies ocean algae the nutrients they need for growth. This is one of the central reasons why warming ocean temperatures pose such a threat to biodiversity-based life-support function.

Lovelock notes that the current climate sensitivity models used by the IPCC are linear programs with respect to $CO_2$ concentration increases. He says that the linear approximation of conditions comprising the larger earth system, as derived from current models, cannot be accurate because they do not take life processes into account. While Lovelock's ideas about the earth's capacity of atmospheric self-regulation are controversial, his views on the critical link between careful monitoring of actual natural circumstances and the reliability of climate models are widely shared within the scientific community around the world.

## ABANDONING STATIONARITY:
## NATURE ALWAYS HAS "PUSHED THE ENVELOPE"

Though it has yet to be fully realized, researchers participating in the linked IP3 and $WC^2N$ initiatives have revolutionized our understanding of climate change in two crucial ways, to which Canadians should pay very close attention.

The first is that new hydroclimatic parameters are emerging as a result of climate warming that have to be understood if we are going to have any hope of determining the kinds of hydroclimatic conditions Canadians are going to have to adapt to in the coming decades. This means that entirely new mathematical descriptions of our cold regions have be developed if we are to be able to accurately predict the effects of climate change on our built environment or on the natural systems upon which that built environment depends for its stability and sustainability.

The second indisputable fact observed by researchers funded under the Canadian Foundation for Climate and Atmospheric Sciences program collectively is that hydrological systems everywhere in Canada have already begun to change. Climate change is no longer theoretical. Its effects are already being observed and clearly measured. While hydrological circumstances have changed only slightly in some areas, these changes are highly pronounced in the cold regions of Canada. The evidence is incontrovertible. Cold matters.

By examining the parameters by which the dynamics of hydrological processes are defined, researchers have made a profound discovery. We simply do not understand all of the hydrological parameters that are operative in natural systems; and many of the parameters we think we have nailed down well enough to employ in models are no longer accurate because natural system dynamics are being altered by rising temperatures. In other words, we are basing our models on what we know about a world that no longer exists.

Something has suddenly changed in our relationship to the world as we once knew it. What IP3 and WC$^2$N researchers are collectively telling us is that the principal concern related to climate warming is not temperature itself, but the effect of rising temperatures on water in its various forms. The fundamental threat that climate change poses relates to what hydrologists call stationarity. Stationarity is the notion that seasonal weather and long-term climate conditions fluctuate within a fixed envelope of relative certainty.

Stationarity implies stability and a relatively high degree of certainty when it comes to predicting and managing the effects of weather and climate on our cities and our agriculture.

The fact that we have determined that natural phenomena fluctuate within a fixed envelope of certainty suggests, for example, that winters will only be so cold and summers so hot. Stationarity suggests that melt from winter snowpacks will always contribute roughly the same amount of water to our rivers and that rivers will rise only so high in spring and fall only so low in autumn. Stationarity suggests we only have to build storm sewers to a certain size because we know from history that rainstorms only last so long and only result in only so much runoff. Stationarity also suggests that lightning will strike only so frequently and that tornadoes will only form in the most extreme conditions of the weather we have come to know and expect.

The research outcomes of IP3 and WC²N researchers demonstrate that hydrological stationarity is no longer a valid characterization of the way climate and various manifestations such as precipitation behave. To appreciate what hydrological stationarity means and the important role it plays in modern life, it is important to understand the central function that water plays in our planet's weather and climate system. Increased temperatures clearly are changing the way water moves through the global hydrological cycle. Precipitation and other climate patterns are now regularly moving outside the ranges we took to be normal in the past.

What this means is that the statistics from the past related to how surface, subsurface and atmospheric water will act under a variety of given circumstances are no longer reliable for use in the future. The problems this creates are manifold. The stationarity associated with those statistics – the notion that natural phenomena fluctuate within a fixed envelope of certainty – is the foundation of risk assessment in engineering upon which we depend for the construction of our buildings, roads, bridges and other infrastructure. Stationarity is also the foundation for determining insurance rates related to risks associated with the protection of our homes and property from fires, floods, tornadoes and hurricanes. Stationarity is also the foundation of the reliable function of the natural ecosystem processes that provide a stable and resilient backdrop to human existence.

Because climate and earth systems are in constant change, however, it can be argued that stationarity in global hydrological conditions has never actually existed at all and that what we have done is establish our own idea of the range of natural climate variability we think exists. We then built our society and the vast infrastructure that supports it around that range. Though we were and continue to be occasionally undone by floods and droughts, the idea of hydrologic stationarity has served us well for hundreds of years. By assuming stationarity within a defined range we were able to create cities in which millions of people live in security; to build water treatment and delivery systems that provide safe drinking water to billions of people; and to construct transportation systems that allow many of us to travel anywhere we may wish to go in the world in a day. But those statistics no longer work and we are confronted with the fact that the future may not resemble the past.

Past changes demonstrate that climate cycles have occurred regularly on time scales ranging from decades to centuries and longer and are most likely to have been caused by oceanic and atmospheric variability and variability in solar intensity. Taken alone, the consequences of this variability for hydrologic cycles imply that climatic patterns are characterized by non-stationarity. While there have been periods of relative stability, there has never been a time in human history when the climate has not been changing to some extent. Ours is not the first civilization to make the terrible mistake of assuming that the climate experienced over a few generations represented a reliable picture of the climatic circumstances that could be expected over the long term. By building our cities and our shared infrastructure to engineering standards based on climate observations of limited duration, and developing our agriculture on the same limited climatic assumptions, we are repeating the fatal mistakes previous civilizations have made. We have built our civilization and then rapidly expanded it based on false assumptions of climate stability. We have failed to fully acknowledge or anticipate the longer-term variability of our global climate.

In the past 50 years we have discovered not only that we as a civilization have been guilty of short-term thinking and planning, but also that this thinking has resulted in the emergence of new threats which are already amplifying and compounding the problem of climate non-stationarity. Since the Industrial Revolution, anthropogenic greenhouse gases have added an additional climate driver. The United Nations Intergovernmental Panel on Climate Change has stated that most of the global warming observed over the 50 years between 1955 and 2005 is attributable to human activities associated with land-use changes and greenhouse gas emissions.

Evidence of both non-stationarity and our failure to adequately account for it can be observed in our failure to forecast the rate of change global warming is causing in the Canadian North. Our forecasts have been wrong because they were based on temperature and atmospheric parameters that assume a degree of hydrologic stationarity that we now know doesn't exist and never has. We are also finding that the same ecosystem depletions and limitations that contributed to the climate change that brought down earlier empires are now appearing globally, threatening all of us, rich and poor alike.

For all intents and purposes our basic reality as we knew

it in the past has been altered. In a very real way, something as fundamental as the gravity of our world has shifted. It is now realized that if we want to be able to predict the future with any kind of reliability, we need a completely new worldview, a new way of understanding the interconnected, interdependent processes of nature, and a new way of relating to planetary systems upon which we depend to make our lives possible and meaningful. While this discovery may appear overwhelming, the research also suggests that the possibility of creating such a revolutionary new world view is within our scientific grasp – if we choose to reach for it.

## DEALING WITH HYDROLOGIC NON-STATIONARITY

We do not as yet have an adequate replacement for stationarity statistics. We have nothing at the moment that works as well as they once did in helping us feed ourselves, design and build our cities and plan for the future. Nor have we committed enough resources to explore new ways of coming to terms with what has become a rapidly sliding scale of change in the fundamental conditions of human existence on this planet. Once again, climate change in itself is only one factor influencing the breakdown of modern stationarity. The central problem is not global warming per se, but its growing influence – in combination with other human activities – on hydroclimatic circumstances.

As the distinguished US hydrologist Dennis Lettenmaier famously argued in a 2008 *Journal of Water Resources Planning & Management* editorial, "Water resources research has been allowed to slide into oblivion over the past 30 years." Dr. Lettenmaier makes an important point. While researchers have begun to grasp the enormity of the challenges that confront us with respect to managing rapidly changing hydrological circumstances, we have wasted 30 years and still can't get past the old way of doing things. Non-stationarity demands sweeping changes in the way we think about water supply. Until we find a new way of substantiating appropriate action in the absence of stationarity, risks will become increasingly difficult to predict or to price.

The problem that Lettenmaier and others have identified is that we cannot create a completely new branch of mathematics and train a new generation of water management experts in this new field of statistical analysis overnight. Until the federal government terminated its funding, it was the goal of the IP3 initiative to create that new branch of mathematics.

The difference between observation and simulation is held to be one of the most important gulfs to be bridged in contemporary science, especially in the domains of water security and climate-change prediction. Without bridging this gap we will not be able to translate what we are observing through direct monitoring in field situations into useful predictions of what might be happening in natural systems in a variety of scenarios in the future. Without a harmonization of modelling results with what natural systems actually do over time, the predictions science makes about the future will not create a sound foundation upon which public policy decisions can reliably be made.

In John Pomeroy's estimation the fact that hydrological processes in Canada's cold regions are out of established equilibrium ought to be very significant to policy makers in government responsible for the orderly allocation of water to the many sectors in our economy that rely on adequate supplies for their stability and sustainability. The loss of hydrological stationarity will void traditional approaches to how we assess risk in the design of buildings, roads, storm sewers and water treatment systems. This change in stationarity will impact oil and gas production, agriculture and transportation. It will undermine the current structure of natural ecosystems, which in turn will affect the composition and dynamics of our atmosphere and reshape our climate. Another consequence of the loss of hydrological stationarity is that new information for water policy related to allocation, conservation and development is required which cannot be provided by analysis of current observations alone. Improved information can only be obtained from the results of coordinated observation and · from prediction systems that incorporate new forms of data assimilation, enhanced observations, improved model development and continuing process research to deal with evolving unknowns.

To respond to hydrologic non-stationarity, Canada needs to invest in monitoring; in the development of new mathematical and statistical techniques to describe parameters more accurately so that we can more completely characterize what is actually happening in nature; and in the development of new models that accurately represent our changing hydrology so that we can more successfully predict the frequency, intensity and duration of extreme weather events such as droughts and floods.

## THE CONSEQUENCES OF LOST STABILITY

At the time of this writing Canadians were not ready to accept the need to make substantive changes to their way of life in the face of a changing global climate. One of the main reasons for this is the low level of public understanding in Canada of how science works and how it arrives at consensus on conclusions. A second reason for resistance to change resides in the fact that most Canadians do not fully appreciate what cold does for them in shaping the Canadian environment and in terms of the stability it provides as a backdrop to all that we do socially and economically.

While some Canadians may welcome some warming associated with a changing climate, they will not welcome the loss of hydroclimate stability that accompanies that warming. Take Toronto, for example, which has endured four once-in-100-years storms in the past 20 years. This means that storms that could be expected once a century, or had a 1 per cent chance of happening in any given year, occurred four times in only two decades. One of these storms caused $550-million in infrastructure damage in just two hours.

The same increase in the frequency of extreme weather events has been observed in other Canadian cities, including Victoria, Vancouver, Edmonton, Calgary, Saskatoon, Winnipeg and Halifax. It is here – with the immediate prospect of more frequent floods and more intense drought, heavier rainfalls and bigger snowstorms – that our climate woes could really begin.

La Niña years, in which cooler temperatures and greater precipitation occur, will allow some people in some places to keep believing that nothing is changing, but those years will become less frequent over time. Researchers at NASA concluded in January 2010 that despite large year-to-year fluctuations associated with the El Niño–La Niña cycle of tropical ocean temperature, the global temperature continued to rise rapidly in the past decade. In the wake of the worst fire season in history in Russia in 2010, the world reached a record high global temperature during the period, as measured by instrument data. Pakistan experienced the fourth-highest temperature ever recorded, 53.7°C, in June 2010.

Environment Canada's lead expert on climate-change adaptation, Ian Burton, coined and defined the term "adaptation deficit," which essentially means not recognizing or fully acknowledging the evidence that our climate system and related services are moving in a direction that is beyond the range of our current capacity to adapt. The graphic on page 160 illustrates the fixed envelope of relative certainty within which we have come to expect our weather and climate to fluctuate over a typical period of time. This is the range of possibilities which, say, engineers would consider in designing major infrastructure such as bridges and highways. The graphic also portrays the increasing number of extremes we can expect to have to cope with. The message is that the loss of stationarity is already making our society increasingly vulnerable to accelerating climate-change effects.

Ever more frequent extreme weather events continue to drive home the realization that the enormous cost of repairing the damage caused by such events could make it very difficult to sustain our prosperity while at the same time protecting and improving our environment.

What we are seeing is that the loss of stationarity is likely to be far more costly than anyone ever projected. Because of our economic circumstances – at the time of this writing – we appear to be in an awkward bind. Governments maintain that our fragile economic recovery cannot be sustained if we pursue climate-change mitigation too rapidly or at all. Nor, because of huge public debt and the increasing costs of extreme events, can we afford to adapt in any meaningful way. The problem is that we can't afford not to, either.

# CHAPTER 6

## Two Degrees of Separation: What a Changing Climate Means to Our Identity as a Nordic People

In the wake of the failure of the Convention of the Parties 15 in Copenhagen in December 2009 to come to agreement on an integrated international solution to the global climate-change threat, many felt that humanity's collective capacity to address the most serious problem facing our civilization had been mortally wounded by competing national self-interest. Only later did the realization present itself that Copenhagen was not the complete failure it was made out to be in many political and environmental circles. Three major advancements were made at COP 15:

- financing was announced to support further efforts to reduce $CO_2$ emissions and fund ongoing negotiations;
- agreement was reached to encourage greater transparency in the reporting of actual emissions; and
- support was granted for the development of a more credible verifiability framework.

The most important advancement made in Copenhagen, however, was consensus among nations that the mean global temperature should not be permitted to rise more than $2°C$.

What is interesting about Arctic warming is that because

much of the permafrost in the North is only 2° below the freezing point, even the limited temperature rise the rest of the world is willing to accept will have catastrophic effects on northern landscapes and livelihoods. There is only two degrees of separation between what the Arctic was and what it will become in a warmer world.

David Livingstone is a former senior official with what at the time was called the federal Department of Indian & Northern Affairs, where he was responsible for environmental issues in the Northwest Territories. Mr. Livingstone is now the chair of a scientific advisory committee that is working to find ways to continue the research initiated under programs funded by the Canadian Foundation for Climate and Atmospheric Sciences. The research outcomes from IP3 and previous studies such as the Mackenzie Global Energy & Water Cycle Experiment have convinced Livingstone that the Mackenzie River basin is one of the linchpins holding North America's water–ice–climate interface together. Like many others who know the North well, Livingstone is concerned that if the stability of this important ecohydrological system is compromised, it could cause the earth's climate to wobble farther out of its current equilibrium, with implications for all the ecosystems on the continent whose stability is coupled to current climate variability.

If you ask Livingstone to describe climate change, the first thing he will say is that in the Canadian North global warming is not in question. Cold, Livingstone explains, is a very good thing in the North, for without it the people who live there cannot maintain their current way of life.

The 2°C global temperature increase limit which the rest of the world agreed in Copenhagen that they would not exceed, Livingstone notes, is already history in the Northwest Territories. Circumpolar Arctic warming is occurring at twice the rate of the global average. The Mackenzie valley in the Northwest Territories, Livingstone explains, is a global hot spot. It has warmed 2°C since the 1940s, and the northern reaches of the basin have warmed even more. The town of Inuvik, located near the mouth of the Mackenzie, near the Arctic Ocean, has already warmed by 3°C.

Livingstone is concerned about the feedbacks that rising temperatures have already created in many parts of northern Canada. From the work of IP3's Phil Marsh and others, Livingstone has been able to recognize these feedbacks immediately:

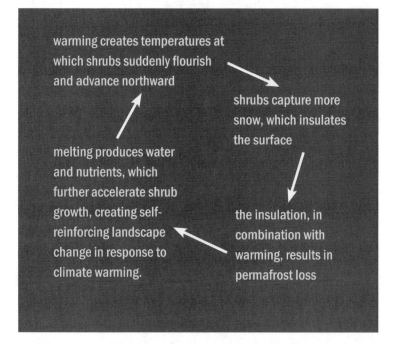

warming creates temperatures at which shrubs suddenly flourish and advance northward

shrubs capture more snow, which insulates the surface

melting produces water and nutrients, which further accelerate shrub growth, creating self-reinforcing landscape change in response to climate warming.

the insulation, in combination with warming, results in permafrost loss

Livingstone points out that, historically, oil and gas operations were permitted to use permafrost as a waste container, and that now that the permafrost is melting, these wastes are being released.

## HOW NORTHERNERS ARE ADAPTING

David Livingstone has witnessed first-hand the very real impacts that climate change is having on the people of the Northwest Territories. Adaptation, he notes, has been both reactive and proactive. As a result, he does not believe adaptation can or should replace the need to reduce greenhouse gas emissions. Precipitation in the Northwest Territories, he reports, has already become variable; wet spring snowstorms are dumping huge snow loads on buildings not designed to hold such weight, and freeze–thaw cycles are more common, demanding more sand and salt on roads and more glycol for de-icing aircraft. On the Mackenzie, floods and low water are affecting barge traffic and ferry service and causing water quality issues related to turbidity, sewage lagoon overflow and tailings pond overflows. Flooding in remote Northwest Territories communities has also become a serious problem.

Livingstone also notes that the winter ice roads so vital for travel and supply logistics in the Northwest Territories have to be built later and closed earlier because of advancing warming. Ice spray technology is being used more widely to create thick, load-bearing ice on frozen rivers and lakes, and ground-penetrating radar has to be used to ensure ice is thick enough to bear the weight of vehicles and to locate problem areas (see bottom photo on page 161).

Livingstone observes that climate change in the circumpolar North also has consequences for the health of the people who live there. He noted that, due in part to climate change and reduced access, there has recently been a shift away from traditional hunting and gathering to more western-style diets. This has led to increased risk of obesity, diabetes and cardiovascular problems among aboriginal peoples, as traditional diets of wild game and fish were healthier than diets that are high in fat, sugar and salt.

Livingstone notes that projections of continued warming that appear in all climate models point to the need for ongoing proactive adaptation. He believes that continuous research and constant reiteration of adaptation strategies will be necessary in northern and Arctic Canada if the people are to have any semblance of the life they led in the past. Livingstone is worried, however, that this research may not keep up with the changes warming has unleashed in the Northwest Territories. He is not sure what northerners will do when the land they are used to standing on melts out from under them. He is concerned that something absolutely fundamental to the northern way of life is under threat. He is not alone is this fear.

## CLIMATE CHANGE AND INDIGENOUS TRADITIONAL KNOWLEDGE

Half of the population of the Northwest Territories is of aboriginal descent. Aboriginal cultures are known to have inhabited the region for thousands of years, relying on the land and its resources to provide food, clothing, water and all the necessities of life, leading to a detailed knowledge of the land, animal behaviour, seasonal and climatic changes and ecological relationships. Aboriginal peoples in the Northwest Territories today maintain strong ties to their traditional ways of life and culture even as they adopt and embrace modern technology and lifestyles. To them, traditional knowledge is what ties them to place and assures the sustainability of their culture and lifeways.

When viewed from the context of the western interpretation of the natural sciences, traditional knowledge is seen to be made up of a logical system of organized knowledge based on empirical data that draws on observations over time, and which historically has been codified and transmitted through oral narrative. Stories told by elders recount their personal experiences or those of their ancestors, while incorporating important information both about environmental elements and processes and about the underlying values or worldview that informs their interpretations of changes in the environment.

Though such knowledge is timeless, the concept of "indigenous knowledge" or "traditional knowledge" has only recently become a part of scientific, academic and regulatory discourse with respect to the governance of water. Such knowledge, however, is increasingly seen as valid and has been shown to play an important role in modern environmental management and decision-making.

The range of domains in which traditional knowledge has proven useful in managing for today and planning for the future of the North is staggering. Traditional knowledge has already proven invaluable in understanding changes in vegetation, fisheries and wildlife populations and movement. Traditional knowledge can also be brought to bear on matters related to air quality, temperature and wind and how climate change may be affecting the location and state of permafrost, soil conditions and terrain stability. In terms of hydrology, traditional knowledge can inform decisions about stream conditions, changes in water temperature and quality, seasonal flow levels, the location of – and changes in – subsurface aquifers, and seasonal ice conditions.

Rapid Arctic warming, however, is undermining traditional ways of knowing and all that they stand for in northern cultures. Traditional knowledge, as linguist and ethnologist Robert Bringhurst reminds us, is really a functioning community of stories that strives to maintain both its coherence and relevance through constant and iterative retelling within a society over time. In this respect it is similar to a living literary canon upon which a society constructs an ideological image of itself that can be subjected to slight but not fundamental changes in its focus and intent in the midst of long-term change. The problem, however, is that the changes that are likely to affect indigenous and local ways of living in the North most profoundly are not those of the incremental kind. It is now realized that the rate and extent to which the northern climate is changing may make even the most enduring mythology difficult to sustain and embrace. If these changes continue to accelerate, the rate and extent of them may well undermine the foundation of the meaning and relevance of the traditional knowledge that currently exists. It may even undermine the possibility of traditional knowledge itself, at least as it is known presently. Should that occur, western science – indeed all contemporary belief systems – will likely also be questioned.

## NORDICITY: COLD IS CULTURAL

For Canadians, one's relationship to cold can be a very personal matter. While northerners clearly understand the extent to which cold matters to their identity and are very open about it, southern Canadians are less inclined to make the connection. For many southerners the fact that Canada is truly a northern country only becomes clear if they are fortunate or adventurous enough to travel in the Arctic. If one has not travelled extensively "north of 60," one may not fully appreciate how cold can work its way into one's psyche and transform one's worldview. One feels more of a Canadian for having experienced directly why cold matters, to us and to the rest of the world. This is evident in our literature. Despite the fact that all of the early literature of the High Arctic was written by English naval officers, this literature is surprisingly varied and rich. The early literature that is particularly meaningful is that which describes life and death situations which, because of the nature of the

Arctic climate, were a regular feature of most early exploration accounts. Among these accounts, the most famous were written by John Franklin.

Between 1819 and 1822 Franklin led a small overland expedition that retraced the route Samuel Hearne had taken between Canada and the shore of the Arctic Ocean on behalf of the Hudson's Bay Company in 1770. It was this expedition, which followed the course of the Coppermine River, which author Rudy Wiebe so magnificently recreated in *A Discovery of Strangers*. One of the central themes of this book was Franklin's almost inhuman arrogance, which was founded on what was for the time a typically British denial of the physical conditions that surrounded him and his men. They continued to live as though the Coppermine River in winter were just England with snow. Europeans like Franklin, Wiebe points out, only wanted to get through the North so they could be somewhere else. But the Arctic, as Wiebe explains, was too much itself to be merely a means for anything. The movement of human beings across the land was analogous, in Wiebe's estimation, to "the line water draws on the land."

In almost all of his writing, Wiebe addresses the largely unconscious northern bias of our culture, our "nordicity" as he calls it. Canadians, Wiebe points out, often ignore their nordicity. Wiebe, however, did not deny his deep existential connections to the snow and ice that shaped and continues to shape his identity. Unlike John Franklin and many southern Canadians today, he did not wish to be somewhere else. Wiebe made it clear that he "desires the true North, not passage to anywhere." He accepted his nordicity and wished other Canadians would do the same.

Wiebe's 1989 book *Playing Dead: A Contemplation Concerning the Arctic* explored the suspended sense of commitment to place so common among Canadians long before a warming Arctic began to force us to re-examine the meaning and value of the nordicity so many Canadians have neglected. Though the value of a book of this intellectual and literary scope cannot be confined by a single quote from it, Wiebe's sense of outrage at our failure to appropriately recognize the influence of the North in our national character is well expressed in the following:

> Perhaps if I had nerve enough to acknowledge it – and I hesitate to say it, such an ancient, old-fashioned "secret" to which one hardly dare confess in the present dark age of irony – I need wisdom. Wisdom to understand why Canadians have so little comprehension of our own nordicity, that we are a northern nation and that, until we grasp imaginatively and realize imaginatively in word, song, image and consciousness that North is both the true nature of our world and also our graspable destiny we will always go whoring after the mocking palm trees and beaches of the Caribbean and Florida and Hawaii; we will always be wishing ourselves something we aren't, always standing staring south across that mockingly invisible border longing for the leeks and onions of our ancient Egyptian nemesis, the United States. Is climate, is weather to be all that determines what we think ourselves to be?

From Rudy Wiebe and other authors we learn that Canadian nordicity is not solely defined by how much of our country is arctic in character but by the extent to which Canadian life is shaped by the persistent presence of snow, lake and river ice, seasonally and permanently frozen ground, glaciers, sea ice and ice caps that together compose our planet's cryosphere. It is the presence of the North in the South that makes Canada and its people so unique. It is the fact that most of us live in or within sight of the world's thin layer of snow and ice that makes us different. Being Canadian means you have accepted the terms and conditions associated with living in an ecology and culture of cold. Our nordicity is something we share, whether we acknowledge it or not, with all other circumpolar peoples.

## COLD ALSO MATTERS AS A WATER-STORAGE MEDIUM

Global hydrological cycles also respond in their own unique way to nordicity. In *The World in 2050: Four Forces Shaping Civilization's Northern Future*, climate scientist Laurence Smith goes to considerable lengths to explain how cold in itself in a reservoir for water. Smith tells us that while a droplet of water in a river may flow down an undammed river in a few days, a similar drop of water bound up in perpetual snow, in the ice of a glacier, within a confined aquifer or a deep ocean current could be confined there for decades, even centuries. But additional water from these other hydrological sources is always quietly entering our streams and rivers, which explains why, despite the fact that the atmosphere only stores about 2000 cubic kilometres of water at any given moment, we are nevertheless able to withdraw almost twice that volume from all the world's rivers at any time.

The overarching lesson Canadians might learn from this is that the availability of these additional volumes of water is a product of a stationarity upon which we can no longer rely. Unlike a glacier or a mountain snowpack, neither the atmosphere nor rivers have storage capacity from which they can draw in dry times, nor can they capture water in wet times to release later. Because their storage capacities are so limited, we are vulnerable to the smallest variations in throughput and highly vulnerable to extremes such as droughts and floods. While we have done much to reduce this vulnerability through the construction of millions of dams, reservoirs, artificial lakes and ponds, Smith notes that we still only have enough of these to artificially impound slightly less than two years of water supply. Without snow we would be in trouble.

Snowpack, snow cover and glacier ice provide free water storage to an extent humanity could never afford to replace. Glaciers and permanent, year-round snowpacks are particularly valuable because they outlast the summer. As Laurence Smith points out, this is valuable because it means glaciers and snowpacks can accumulate extra water during cool, wet years when agriculture is likely to need it least, and release water in warm, dry years when farmers and others need it most. This is exactly what happened under the stationarity to which we have become accustomed. But that stationarity no longer exists.

Rising temperatures are already reducing the benefits of natural storage by increasing the amount of rain that falls in winter and by shifting the spring melt to earlier in the season. Because the growing season is determined not only by temperature but by length of daylight, farmers in many areas will not be able to take advantage of earlier melt by planting earlier. In addition they are likely to face longer, hotter summers, which will mean less water being available in late summer when it is needed most. Herein resides a huge problem associated with our failure in Canada to understand the significance of altered stationarity and our lingering and persistent attachment to outmoded ways of thinking and managing water.

Smith emphatically argues, however, that no amount of engineering will be able to replace the storage that formerly was provided to us free by lingering snowpacks and glaciers. Under the current and increasingly unpredictable terms of non-stationarity, even if we quadrupled the number of reservoirs worldwide, we would never come close to replacing natural storage. And even if we did, we would still end up with less water, because unlike snow and ice, liquid water evaporates quickly and in huge volumes from open reservoirs. Our greatest challenge now is to adapt to a shrinking water-storage capacity that was once provided to us for free by annual snow cover and ice.

We can't hold all of our mountain meltwater back, as Smith points out. There will be less water in many of the mountain ranges of the world, and more of what is left is leaving upland sources to run into the sea. It is the hydrological throughput that matters, so even where there are abundant dams and large reservoirs, as in places like the US Southwest, there will not enough water to fill them under reduced regimes of snowpack and snow cover.

Smith makes it very clear he is not confident that the kinds of benefits brought about by a warming world are going to compensate in any way for the loss of what we presently have, especially in the Arctic. He notes, for example, that because of the loss of hydrological stationarity, it is an oversimplification to suggest that Russian and Canadian agriculture will flourish as a result of warmer mean temperatures in both countries. The most expansive agricultural regions of both countries lie on what are essentially dry plains. Given that such projections are premised on a stationarity that no longer exists, there can be no guarantee that gains in agricultural productivity in the northern reaches of these countries will be adequate to make up for losses elsewhere because of increasing temperatures and changing hydrological regimes in the south. In Canada there is also the problem of soil types. While prairie soils are ideal for growing grains, much of the north into which agriculture would have to expand to increase productivity is underlain by less productive soils and by the hard rock of the Canadian Shield.

Because our risk assessments were made under the out-moded 20th century mentality of stationarity, hydroclimatic considerations related to our nordicity were not fully taken into account when governments decided on our behalf that Canadians were going to simply adapt to climate change rather than concentrating in a timely manner on effective mitigation. The rapid meltdown of Arctic ecosystems that is taking place right in front of our very eyes indicates that this may have been a very grave mistake indeed.

## SCIENCE TO THE RESCUE

Through research conducted by scientific networks, including IP3, we have discovered that Canada's actual population and greenhouse emissions are less relevant in terms of climate change effects than the size of Canada's land mass and the disproportionate impacts we are collectively having on it and on the oceans that surround us. Higher temperatures are occurring in tandem with a number of other human-induced changes that threaten to cast everything we thought we knew about the world into limbo. These actions fall generally under categories of change that are eliminating parameters of natural processes before we have even begun to fully understand them. These impacts include habitat destruction, habitat fragmentation, overharvesting (particularly relating to oceans) and the active transfer of invasive species from one ecosystem to another. These impacts in combination result in species extinctions which have the effect of cascading through ecosystems in ways that obliterate known and as yet undiscovered parameters that once contributed to the stationarity upon which we constructed our view of how much of humanity the natural world could stand without collapsing.

Northerners know that any further warming may make their nordic way of life impossible. Because hydroclimatic change is evident all around them, they know their identity – their very nordicity – is being threatened. As a consequence, northerners take both traditional and local knowledge and western science very seriously. The work of IP3 and other initiatives funded by the Canadian Foundation for Climate and Atmospheric Science matters in places like Whitehorse, Yellowknife and Iqaluit. In these locales the results of scientific research related to climate and water are too immediately important to be ignored.

## WHEN DEMOCRATIC GOVERNMENT ACTUALLY GOVERNS: MITIGATING CLIMATE CHANGE WITH SOUND WATER POLICY IN NWT

The region of the North that has responded most quickly and effectively to the findings of CFCAS-funded research is the Northwest Territories. Through a great deal of serious collaboration the people of NWT decided that though there were many issues they disagreed about, everyone was concerned about water. They were concerned that climate change would affect not just the volume of water available in the Mackenzie basin, but its temperature and therefore its utility. Because there was less debate about water than about many other resources, people began to see water policy reform as a vehicle for addressing not just climate change but a broader range of environmental and economic concerns.

After two years of dialogue, representatives of more than 30 communities agreed that the most effective way of quickly defusing the climate threat was to manage water more effectively and in a more integrated way. They believe that better water management in itself will make it possible to solve a number of other problems while at the same time assuring that climate-related water quality and quantity issues don't threaten their economic and social future.

Under the leadership of deputy premier Michael Miltenberger, the people of the Northwest Territories determined that a new territorial water ethic could be a means of ultimately achieving greater capacity to adapt to climate change while generating a great many other lasting social, economic and environmental benefits along the way. They then articulated and affirmed that ethic.

In May 2010 the government of the Northwest Territories, in partnership with Indian & Northern Affairs Canada (now called Aboriginal Affairs & Northern Development Canada) released Northern Voices, Northern Waters: The NWT Water Stewardship Strategy. The strategy seeks to protect aquatic ecosystems, ensure safe and reliable sources of water for the residents of the Northwest Territories and sustain traditional ways of life in the North in the face of rapid hydroclimatic change. Amazingly, this strategy demonstrates that action to mitigate climate impacts on water can be a positive rather than a negative stimulus for the economy and can be something that protects and enhances rather than diminishes local quality of life.

The elite independent national Forum for Leadership on Water, called FLOW, was invited by the NWT government to review the stewardship strategy to determine its strengths and weaknesses and establish the value their approach to managing water might have in southern Canada. The members of FLOW identified a number of principles upon which the NWT strategy is founded that they deemed to be superior singularly or in tandem with principles underpinning other provincial or territorial water strategies created to deal with hydroclimatic changes in the coming decades. These strengths fall into four main areas:

- fundamental strengths

- strengths associated with an ecosystem approach to water governance
- strengths associated with the need for sound science and accurate information as a foundation for decision making
- strengths of the strategy associated with models of co-governance of water in the Northwest Territories.

The first thing that impressed the members of the forum was that the governments of the Northwest Territories and Canada did what governments are supposed to do when faced with a matter as serious as that posed by rapid hydroclimatic change: they governed. The provincial government and its federal and local partners assumed timely, complete and proactive responsibility for broad community collaboration leading to the development of a fully integrated new watershed-based territorial water stewardship strategy.

Northern Voices, Northern Waters is a precedent-setting document nationally and globally because of the way it was crafted and because of the degree of accountability and transparency it demands of everyone involved in its implementation. The agencies responsible for the development of the strategy took seriously their fiduciary and statutory responsibilities for the management of water and watersheds within their jurisdiction. They did not back away from the project because water policy reform on this scale was too difficult to agree on or to legally orchestrate or too politically sensitive to address, as so often happens in the south. They did not bow behind closed doors to individual sectoral interests or hold back on the extent of reform in response to special pleading. They did not create a strategy that focuses on economic opportunity associated with dealing with symptoms of water supply and quality concerns while ignoring the fundamental environmental issues that are so often the source of these problems.

*In Buenos Aires in November 2010, Northwest Territories deputy premier J. Michael Miltenberger addresses the Seventh Biennial Rosenberg International Forum on Water Policy on the subject of the Northern Voices, Northern Waters water stewardship strategy.*

Photograph courtesy of R.W. Sandford UN Water for Life Decade, Canada

What they created is not half a strategy that avoids serious weaknesses in current water governance, such as those related to already established water rights, allocation privileges or water policy precedents. In the estimation of the Forum for Leadership on Water, the NWT government and the federal ministry of Aboriginal Affairs & Northern Development acted responsibly in collaborating with affected constituencies to find and implement solutions based on sound science that work for all of society over the long term. They did what was necessary, not what was easy. In other words, they got the whole job done, not just the parts of it that would be an easy political sell.

What the NWT produced was complete water policy reform

– a thorough overhaul of governance structures as they relate to the management of water and the land–water nexus in the Mackenzie basin and beyond. The Forum for Leadership on Water observed that the fact that a territorial government in northern Canada had the courage and the wherewithal to initiate such broad but necessary reform in the face of rapid hydroclimatic change should in itself be an inspiration to others.

Citing the work of IP3 researchers Bill Quinton and others on issues such as permafrost loss and ecosystem change, the NWT government deemed it unwise to base future development in the region on natural resources material and energy throughput only. The goal of the current government is to create a post-resource-extraction economy. The creation of such an economy will require a shift of focus and priority to the protection of land, water and animals together first, with resource development occurring within the constraints of larger ecosystem protection strategies. The object will be to use the capital accrued through resource extraction to meet the objectives of the water strategy and maintain economic health.

## FLOW'S RECOMMENDATIONS FOR EVEN FURTHER POLICY IMPROVEMENT

The Forum for Leadership on Water supported the governments' goals of conservation of intact freshwater ecosystems, decision making that places nature's needs for water at the forefront, the establishment of ecological sustainability boundaries, and investment in ecosystem restoration, not just in the Northwest Territories but throughout the Mackenzie basin, which also includes British Columbia, Alberta and part of Saskatchewan. While the members of FLOW supported the notion that water for nature is water for people, they pointed out that to achieve the goals put forward in the Northern Voices, Northern Waters strategy, it will be necessary to practise expanded and continuous monitoring in order to clearly determine how much water remains for nature and how much can be appropriated for human uses, right down to the site-specific level of every development proposal.

It is widely held that you can't manage what you haven't measured. Improved surface and groundwater monitoring and mapping are central to water security regionally, nationally and globally. Effective monitoring and ongoing research will allow the NWT to track and measure changes in water quality, quantity, rates of flow, and biological parameters over time and space, determine what may have caused these changes and whether modifications have to be made in the management of human activities within the watershed or beyond.

The forum noted, however, that although monitoring knowledge and expertise exists, it is not adequately supported. Members of FLOW expressed fear that the absence of relevant monitoring information will have a cascading effect over time on the success of the proposed strategy. In order to manage for the future, the NWT government will have to rely on predictive models, and long-term trend analyses require long-term data. No model can recover data from a single lost year. Further complicating this process is the fact that the ecological baseline in the Northwest Territories is shifting rapidly as a result of climate warming and economic development. Water and climate models are only as good as the information that powers them. If the results of models are different from what years of common sense garnered from traditional and local knowledge indicate is happening, the models will not be trusted, which could cast doubt on the effectiveness of the Northern Voices, Northern Waters strategy and diminish support for it.

One way of addressing this problem, FLOW recommended, is to work with other provinces and territories to reaffirm the importance of the rigorous framework of consistent and regular hydrologic and water-quality monitoring already established within Environment Canada through the National Water Survey and its provincial and territorial partners to provide a permanently supported baseline of information in support of ongoing assessment of the effects of a changing climate on water security.

The members of the Forum for Leadership on Water recommended that the government of the Northwest Territories, even given limited funds, invest in data collection as a first priority in the implementation of the Northern Voices, Northern Waters strategy. Breakthroughs in contemporary technology can make improved monitoring affordable. In order to enhance data analysis opportunities, monitoring data should be easily accessible to outside partners via portals such as the Internet. The results of outside data analyses will provide the territorial government with valuable information and insights that will help guide implementation of the water strategy.

The members of FLOW thought there were only two really serious weaknesses in the Northern Voices, Northern Waters strategy.

The first was that accelerating climate change could undermine the implementation and outcomes of the strategy. The forum noted, as others have, that the Canadian Arctic is warming faster than anywhere else in the country. It is not just temperature rise that is causing concern, however. It is what temperature changes are doing to the northern hydrology. The hydrology of the Northwest Territories is changing rapidly and all signs indicate that these changes will increase in intensity and frequency as time goes on. Rapid, continued warming in the Mackenzie basin, FLOW members pointed out, could undermine ecological processes to such an extent that the fundamental principles of the Northern Voices, Northern

Waters strategy no longer pertain. If this happens it is not clear to what extent such changes would influence ecosystems in jurisdictions upstream from the Northwest Territories, but effects would be likely throughout the basin.

There are also existing health threats related to climate warming that could undermine the outcomes of the strategy. Modern chemicals pose much greater challenges than more conventional forms of pollution because their sources are often global rather than local; they are invisible rather than visible; and in many cases their effects are chronic and long-term rather than acute and immediate. A very recent example is the discovery of extremely elevated levels of mercury and banned substances such as PCBS and DDT in some fish in northern Canada, with concurrent risks to citizens who consume those fish. Although the exact cause is uncertain, one possible explanation may be that as temperatures in the Arctic rise, snow and ice cover diminish, leading to a profusion of algae, zooplankton and other microscopic life which absorb such pollutants from water and sediments before being in turn consumed by fish.

The members of the Forum for Leadership on Water advised the NWT government to create a carefully orchestrated, partnership-based public education campaign in southern Canada to advance Canadian understanding of the importance of the North to the future of the country. One of the key messages FLOW recommended for such a campaign was that there exists a unique opportunity in the Northwest Territories to finally get integrated water and land management right. The NWT has demonstrated that it is possible to preserve essential ecological elements that are at the foundation of ecological integrity at a time when a changing climate makes it absolutely necessary to do so. If we are lucky, the Northwest Territories may be able to encourage Canadians to see that the climate-related threats that are so dramatically altering the north may soon appear where they live in the south and that the best way to deal with these threats and adapt to change is to carefully manage and protect all the elements of the hydrological cycle that are within local control. Reliable science must be the foundation of all such efforts if they are to succeed.

# CHAPTER 7

## Choosing Our Path:
## The Next Generation of
## Canadian Cold Regions Science

*"The models are already improving and predictive
capacity is advancing through the synthesis
of IP3 and WC²N network experiences."*
—DR. JOHN POMEROY

A few months before the federal government funding for the
Canadian Foundation for Atmospheric & Climate Sciences
ran out, more than a hundred of the best hydrologists, glaciologists and climate modellers in the world met in Lake Louise,
Alberta, for four days to share the outcomes of the projects
they had been engaged in during the previous three years. All
of these scientists were members of one or both of two CFCAS-
funded Canadian scientific research projects: the Improved
Processes Parameterization & Prediction Network and the
Western Canadian Cryospheric Network. The conference
was chaired by the respective principal investigators for IP3
and WC²N, John Pomeroy of the University of Saskatchewan
and Brian Menounos of the University of Northern British
Columbia.

Professor Pomeroy explained that a central finding of the

IP3 project was that Canada's hydrological systems, at least in the country's cold regions, are on the move. The natural variability in rain and snowfall, river flows, lake levels and the area of the country in which permafrost persists are all changing. In scientific terms Canada is experiencing a loss of hydrological stationarity. As a result, precipitation and river flows will be different from what we have come to expect. New ranges of variability will emerge. There will be more and more times when that variability will be outside the range for which our urban and rural infrastructure was designed to function. There will be more times when climate variability will be outside our current ability to adapt.

In Pomeroy's estimation the fact that hydrological processes in Canada's cold regions are out of established equilibrium ought to be very significant to government policy makers responsible for the orderly allocation of water to the many sectors in our economy that rely on adequate supplies for their stability and sustainability. The loss of hydrological stationarity will void traditional approaches to how we assess risk in the design of buildings, roads, storm sewers and water treatment systems. This change in stationarity will impact agricultural production and affect transportation. It will undermine the current structure of natural ecosystems, which in turn will affect the composition and dynamics of our atmosphere and reshape our climate. Another consequence of the loss of hydrological stationarity is that new information for water policy related to allocation, conservation and development is required that cannot be provided by analysis of current observations alone. Improved information can only be obtained from the results of coordinated observation and prediction systems that incorporate new forms of data assimilation, enhanced observation, improved model development and continuing process research to deal with evolving unknowns.

The presentations that followed by IP3 researchers demonstrated how much consideration must be given and how much work must be done just to establish a single scientific fact upon which even the most elementary predictions can be founded. Though it may be laborious, the gradual, persistent improvement of ideas and procedures has proven itself time and again to be a powerful mechanism for the advancement of human knowledge. The rigour of the process should in itself give the public confidence in the scientific method and in the research outcomes of networks like IP3 and the WC²N.

The importance of research is easy to underestimate. At present our global climate models are too crude to tell us what is going to happen specifically as a result of climate change in any given region. They can predict general trends on a very large scale but they cannot tell us what will happen, for example, in a particular mountain range or river basin. In order to improve the accuracy of these predictions, more has to be known about specific conditions and trends at the regional level. That is also what all the researchers in the Western Canadian Cryospheric Network have been working toward. What was interesting about this meeting, however, was that government and private-sector partners all agreed that they were deriving important information from these networks which they were either applying directly to the management of water in specific basins or considering generally within the context of evolving larger policies related to long-term water security.

When representatives of the Cold Regions Hydrology Users Group, an informal committee comprised of agencies interested in the research findings of IP3, were asked if they felt there was a need to continue this work, they responded very positively. They noted that while research programs such as those undertaken by members of the Canadian Water

Resources Association and the Canadian Geophysical Union Hydrology Section generated very useful results, the venues at which those results were presented were too large and too limited in duration and location to provide the dialogue they needed to derive optimal benefit from the findings. They liked the IP3–WC²N model because it allowed greater flexibility for smaller network linkages and the ability to reach out to a broader stakeholder base. They also liked the scale of collaboration offered by IP3 and WC²N because it provided a means for professionals from such widely diverse organizations as government, industry and academia to collaborate easily, with crossover between disciplines.

The members of the Cold Regions Hydrology Users Group also indicated there were additional projects within the cold regions hydrology theme that they would like to have had an opportunity to undertake. These included the opportunity to use the cold regions hydrological model in various small-scale modelling projects in small basin studies focused on specific water management issues. The user's group also wanted to continue developing modelling tools, specifically in areas related to regional evaporation, transpiration and changes in the rates of snow sublimation. There was considerable support for continuing current research at Marmot Creek in the Kananaskis region of the Rockies, as this research provided outcomes that were central to the development of a clear picture of water security throughout all of western Canada in the future. The user's group also noted the value of providing more graduate student fellowships through the Natural Sciences & Engineering Research Council for young scientists who wanted to use the cold regions hydrological model, particularly in the Arctic.

When asked if scientific findings, research products and models developed through IP3 were useful to their organization, representatives of the Cold Regions Hydrology Users Group were unanimous in their support of the need to employ new tools such as the cold regions hydrology model and to market new knowledge within their own organizations. They also indicated that glacier–hydrology–climate modelling was generating results that were critical to their organizations in their efforts to develop meaningful climate-change mitigation and adaptation strategies. Though they expected the number of monitoring sites to decrease in the face of continuing budget cuts, they recognized the crucial importance of comprehensive monitoring to the improvement of hydrologic forecasts and long-term prediction. All expressed a desire to have ongoing access to the data collected by IP3 and WC²N researchers. All were also prepared to supply data and metadata to a common database and archive.

## THE VALUE OF THE SCIENCE TO THE PRIVATE SECTOR: THE BC HYDRO BUSINESS CASE

The value of IP3 and WC²N research to the private sector in Canada was perhaps best articulated in a presentation at the IP3 conference in Lake Louise by Frank Weber of BC Hydro. Weber commented on the direct value of IP3 and WC²N research to a large water utility whose very survival depended upon accurate, up-to-the-moment monitoring of streams and weather; ongoing understanding of changes in mountain snowpack; the capacity to predict the effects of extreme weather events such as floods and droughts; and careful management of reservoir levels.

The subject of Weber's presentation was the use of glaciological data for runoff forecasting. Weber began by outlining that his presentation was meant to serve several purposes. The first was to express his gratitude to the Western Canadian Cryospheric Network for the data and information the researchers in the group have been able to produce to date. This was an important statement, considering that BC Hydro and utilities like it are important partners in research projects of this nature. If the outcomes of major research projects like IP3 and WC²N do not have value to those who rely on better methods of understanding the country's hydrology, then such projects are not likely to be funded in the future. Finally, Weber wanted to pose a number of scientific questions that might become the basis of further research.

He went on to point out that BC Hydro is a power utility with operations and revenues based largely on river flows. Climate variability and change can ultimately affect every aspect of BC Hydro's business. Weber noted that the WC²N was uniquely situated to be of value to the company because of its considerable expertise in cryospheric sciences and, for example, its capacity to model coupled glacier–streamflow responses over time. Much of the data produced by WC²N has been of immense value to the hydrologic modelling activities in which BC Hydro is continuously engaged. Hydrologic models are employed in making short- and long-range forecasts which enable planners to optimize reservoir operations to meet BC Hydro's triple bottom line. These commitments are substantial. BC Hydro serves about 95 per cent of British Columbia's electrical power needs, nearly all of it through hydroelectric generation. The utility operates 80 generating units and 75 dams at 41 dam sites located throughout British Columbia. It also has three thermal generation plants and several diesel units. BC Hydro also distributes energy from a growing number of run-of-river and wind-power generating sites located all over the province.

Frank Weber's job is to provide accurate hydrologic forecasts to BC Hydro planners in a timely manner. This is an enormous responsibility in that accurate inflow forecasts are crucial to reservoir operation. Weber pointed out that if you don't know how much water you are going to have, you don't know how much to store; and if you don't know how much to store you can make mistakes in optimizing the production of your wares, which could dramatically affect your revenues, your environmental footprint and your impact on other water users. It is for this reason that BC Hydro pays a lot of attention to runoff forecasts.

Weber and his colleagues expressed great interest in the findings of the research conducted at the University of Northern British Columbia as part of the WC²N 2005 Landsat-based glacier-extent study which was published by Tobias Bolch and colleagues in 2008. This research, in combination with more precise calibrations developed for the UBC Watershed Model, which BC Hydro also relies on in making its forecasts, is allowing predictions to slowly become more consistent with what is actually observed each year in each of the basins in which the utility operates dams.

In watersheds with glacier cover, one of the major challenges modellers are faced with when setting up process-based hydrologic models is accurate separation of the snowmelt signal from the glacier signal. To aid the correct apportionment of these runoff components, two variables are of particular interest and can be obtained from long-term glacier monitoring programs: the mass-balance equilibrium line altitude, or ELA; and the annual glacier mass balance.

It was very interesting to note the extent to which BC Hydro relies on estimates of equilibrium line altitude for calibrations of process-based hydrologic models. It was even more interesting to discover that the ELA at Peyto Glacier, which lies outside and to the east of the utility's Upper Columbia River watersheds, had to be used to constrain UBC Watershed Model parameters that affect the simulated ELA in the Mica basin, due to the paucity of observations within that basin.

Weber went on to offer empirical "back of the envelope" estimations of the annual total glacial runoff contribution in basins crucial to BC Hydro operations. These calculations, he noted, were all based on Peyto Glacier net annual mass balance data collected, interpreted and shared by Mike Demuth and his associates at the Geological Survey of Canada. During the hydrologic model calibration process, BC Hydro modellers used these empirical "back of the envelope" glacier runoff estimates as independent checks for estimates produced by the UBC Watershed Model. In the local drainage basins that contributed water to the reservoirs behind Mica, Revelstoke and Arrow Lakes dams, Weber estimated the annual glacier contribution to be currently about 8 per cent, 6 per cent and 2 per cent of total inflow, respectively.

Weber made it very clear that BC Hydro relied heavily on the glacier mass balance observations made on the Wapta Icefield by Mike Demuth and Steve Bertollo and the Glaciology Section of the Geological Survey of Canada. For watersheds located in the Coast Mountains of British Columbia, BC Hydro also relied on similar mass balance studies undertaken by Dan Moore and Mike Demuth on Place Glacier and on the developing dataset that formed the ongoing State and Evolution of Canada's Glaciers program, also developed by the Geological Survey of Canada. Weber's point was that hydrologists have little data to accurately characterize glacier melt in basins where no glacier monitoring program exists. This results in high uncertainty with respect to the accuracy of glacier and snowmelt simulations, which can, in extreme years, translate into inaccurate river flow forecasts. For this reason, BC Hydro is very interested in the ongoing monitoring of changes in the area and volume of existing glaciers.

Weber observed that progress toward improved parameterization and prediction is essential to his company's and the country's future. The implications of the conditions that were emerging as a result of the loss of glacier ice in the mountain West were simply too significant to ignore.

While hydrologists are typically focused on communicating the numerical results from their empirical studies and model predictions to their audience, it is equally important for them to support their findings with a congruent story line. For example, to illustrate the complicated and sometimes controversial topic of climate change, Weber demonstrated the value of photos "that speak a thousand words" to drive a message home quite effectively: in this case, a pair of images illustrating the extent of change in the length and volume of Robson Glacier, located near the Mica headwaters.

Weber outlined the potential impacts that further glacier loss would have on hydroelectric generation. All other things being equal, these impacts might include reduced reservoir inflows, which would translate into lower generation capacity, which would in turn result in diminished BC Hydro revenues. Impacts could also include a more frequent need to use reservoir storage to maintain flows required for aquatic ecosystem health and healthy fish populations at the cost of forgoing power-generation profits. The threats also include the decreased buffering of inter-annual flow variability.

Weber concluded by listing a number of core scientific questions BC Hydro felt should be posed in future hydrologic research in the mountain West. In his estimation, hydrologic forecasting efforts of British Columbia's major power utility would greatly benefit from the cold regions research community providing more and better information about the spatial extent and magnitude of seasonal snowpack. He also indicated that BC Hydro is seeking to develop novel, robust and mobile technologies that will provide site-specific, real-time measurement of snow-water equivalent. Such tools, Weber asserted, might emerge through the further development of cost-effective new techniques for remote sensing of snow-water equivalent, associated perhaps with breakthroughs in satellite technology. BC Hydro also supports the ongoing monitoring of glacier extent and mass balance studies such as those undertaken over the long term by the Geological Survey of Canada and those completed more recently by research teams involved in the work of $WC^2N$.

Weber also underscored the importance of continuing efforts to improve modelling techniques related to glacier mass balance, and the need for hydrological models that can accurately simulate the non-stationary effects of changes in landforms caused by glacial recession and changes in forest cover. Weber also wanted the research community to expand the work to derive improved algorithms for rain-on-snow events that would help BC Hydro generate better flood forecasts.

In Frank Weber's mind, the work undertaken by the IP3 and $WC^2N$ networks was just the beginning. In tandem these advances in science would allow Weber and his runoff forecasting team to improve short- and long-range seasonal inflow forecasting as well as better quantitative assessment of net impacts of coupled climate and land cover changes on reservoir inflows. Far more needs to be done, he said, and BC Hydro wants to participate.

## THE SPIN-OFF BENEFITS OF GOOD RESEARCH

While vast improvements have been made in the identification and characterization of the natural processes that define the hydrology of the mountain West, ongoing investment in further research is necessary if current models are to be able to offer reliable predictions of how much water will be available in the Canadian West under future climate scenarios and how ecological conditions in this region will respond to changing hydrological circumstances. Beyond the generation of crucially important hydrological and climatic predictions, another benefit of continuing this research is that it generates innovative new technologies that may have broader practical and commercial applications.

John Pomeroy and his colleagues, for example, have been exploring the possibility of determining snow-water equivalent through the use of an acoustic wave to determine how much liquid water exists in the snowpack and in drifts. This experimental apparatus has been successfully tested at sites in Saskatchewan, the Yukon and the Rocky Mountains. Such technology, if perfected and miniaturized, could have other important applications. Imagine a wristwatch-sized computer that could depict the layers in the snow beneath a backcountry skier or snowmobiler in steep mountain terrain. Conceivably such a device could instantly warn of avalanche risk without the need to dig a snow pit to determine the presence of unstable layers within the snowpack. These are the kinds of practical spinoff benefits that often result from pure research.

## CONTINUING THE WORK OF THE IP3 & WC²N RESEARCH NETWORKS

When Erica Wilson Butcher spoke at the Lake Louise conference on behalf of the Canadian Foundation for Climate and Atmospheric Sciences, she announced what everyone already knew: that her organization had been denied further funding by the current federal government. This meant that the Canadian Foundation for Climate and Atmospheric Sciences would not exist beyond its founding iteration, under the aegis of which it had it had funded the research activities of more than 100 investigators in the IP3 and WC²N networks.

Butcher talked about the importance of user groups to the CFCAS research process. The mere mention of the importance of people using the research outcomes of these two networks puts into relief the deep flaws in this system. Just when government agencies, electrical utilities, industry, agriculture, municipalities, tourism and the average citizen want more information upon which to base their adaptive responses to climate change, it won't be there. Interested parties may be able to access information through some of the individual researchers, but the networks of coherent discourse and information-sharing will no longer be funded. A very strong case can be made, however, for continuing to support the research efforts of these two important science networks.

## THE RELEVANCE OF IP3 AND WC²N RESEARCH OUTCOMES

The first question policy makers might ask would relate to the practical relevance of the IP3 and WC²N research outcomes. How do these outcomes contribute to addressing real problems Canada is facing with respect to the management of our freshwater resources? The collective research outcomes from the two networks are useful in all of the dozen problem domains that have been outlined.

These research outcomes blow the myth of limitless abundance out of the water, so to speak. Given current growth trends, significant parts of the country are going to be water scarce in the future if they are not already. The decline of streamflows in Alberta's major rivers, for example, has clear public policy implications, especially where streamflow is already fully allocated at its current level. The volume of water available to the City of Calgary and to downstream irrigators and prairie communities will diminish. There will be less water available for petroleum upgrading and industrial and other activities in Edmonton. There will be less water available for oil sands activities. There will be less water in both the Peace and Athabasca systems, which will have implications for aquatic ecosystems in the Peace–Athabasca delta and downstream in the Slave and Mackenzie systems.

The research proves the great value and urgency of accurate monitoring as the foundation of forecasting and prediction. It also cautions that the verification of simulations against careful field observations is critical to the development of reliable models.

The research outcomes paint a clear picture of how energy generation in the future could be dramatically affected by water availability. Clearly, large-scale mining operations like the oil sands are going to be affected by changes in hydrologic regimes. This suggests urgency in developing new technologies that reduce the dependence of these operations on large volumes of water. Researchers like Sean Carey are already leaders in this field.

These research outcomes also suggest that there are going to be many more of us competing for much less water in the West in the future, and that to avoid conflict we shouldn't wait another hundred years to characterize First Nations and other water allocation rights. We should characterize them now in full knowledge that hydrological regimes will be different in the future. This means that reform of outmoded first-in-time, first-in-right water allocation doctrines may not be optional.

The collective research of the IP3 and WC²N networks clearly indicates that we have to get moving on the improvement of agricultural practices or the damage these practices do to our water resources will be exacerbated by reduced flows in many systems.

We also have to recognize nature's own need for water. If experience abroad is any indication, Canada would be wise to move forward toward supplying water to nature before over-allocation for human purposes makes it difficult if not impossible to do so.

With growing pressure on ever-more-limited supplies, municipal and industrial water reuse is no longer an option. So we had better get on with it.

The research outcomes of these two networks indicates that it is not just urban water infrastructure that is vulnerable to more extreme climate variability – all infrastructure is vulnerable, including roads, pipelines, power lines, everything. And that vulnerability will only grow.

The research outcomes of the two networks indicate that it is unwise to ignore the effects climate change will have on our society. This realization in itself suggests the urgency of keeping Canadian Foundation for Climate and Atmospheric Science and these two research networks going.

Finally, this research clearly indicates that there have been few times in human history when we have needed effective government more than we do now. Our sustainability depends on it. Sustainable governance can begin with water. From a public policy point of view, perhaps the first thing that may need to be realized is that the sheer number of unresolved issues we face with respect to the management of our water resources makes us all highly vulnerable to the inevitable effects of deep and prolonged drought and climate-change-related impacts on our water supply. If our supply of water in our already fully allocated rivers were suddenly reduced, the number and complexity of the problems that have already emerged would overwhelm us on all levels, just as occurred recently during the prolonged drought in Australia. It is therefore important to seriously address, as quickly as possible in the near term, as many as possible of the problems we have identified. The object should be to deal with the keystone problems immediately so that when water scarcity or drought inevitably arrives, it can be managed. These problems cannot be resolved all at once, but they can be resolved one or two at a time.

Fortunately there is still room to move in Canada with respect to water resource management. Dealing with converging water supply and water quality issues is an important rehearsal for dealing with other troubling convergences that are occurring at the interface between humans and the global environment upon which we depend for economic and social stability.

## TOWARD THE FUTURE

John Pomeroy began his summation of the Lake Louise conference and discussion about future directions for the IP3 and WC²N research networks by underscoring what was already obvious to everyone in the room. The problems associated with properly characterizing and modelling the effects that changing land use and a warming climate will have on the hydrology of the mountain West are not going to go away anytime soon. In Pomeroy's estimation, researchers and the user groups that relied on their outcomes have no choice but to carry on. Even if it is not possible to find further funding to keep this excellent team of researchers together, Pomeroy said, it was of great scientific importance that the investigators that were part of these two networks keep collaborating even after the network funding expires.

Professor Pomeroy then indicated that he had in no way given up seeking funding to keep this fine group of researchers and their private-sector partners working together. He noted that stresses were beginning to appear in the institutions mandated to manage water resources that would ensure that the value of the research undertaken by IP3 and WC²N – and their sister network the Drought Research Initiative – would have to be recognized. One of the major institutions he worried aloud about was the Prairie Provinces Water Apportionment Agreement, which he predicted would fail within the next few years, potentially sparking a divisive free-for-all for water resources among the western provinces.

Pomeroy reminded those in attendance that the research community was not without good ideas on how ongoing research might be organized. He then announced once again his interests in evolving the IP3 and WC²N networks into their next iteration by creating what he called the Coldwater Science Collaborative.

He went on to describe how such a collaborative could be created and how it might be made to aid in the integration of observations and prediction of surface water, groundwater, snow and ice by working up from a series of well-instrumented small research basins in the headwaters to strategically instrumented large watersheds. Pomeroy then demonstrated how broader interagency collaboration could result in the integrated use of water and weather observing stations, intensive research sites, regional atmospheric models, remote sensing, physically based hydrological models and a focus on cold regions to develop the tools to scale predictions up to key watersheds in western and northern Canada.

Finally, it was agreed that an energized new collaboration of existing expertise and institutional capacity would allow participants to collectively address complex water stewardship issues at a higher level of information integration. This could be accomplished by joining together the several, currently fragmented and time-limited western and northern water and climate science initiatives that are presently collecting data or undertaking science for the public good.

Pomeroy argued that while IP3 may have been all about

creating new tools, that function is nearly completed. In the next iteration of the IP3 and WC$^2$N collaboration the goal would have to be to employ these tools. This will mean building monitoring networks for specific purposes, which will require even more collaboration with industrial partners such as BC Hydro and others, whose information needs could be met through the utilization of emerging new modelling tools of the kind that had been demonstrated as being possible through the work of the networks that had been funded by the Canadian Foundation for Climate and Atmospheric Sciences. It will also mean, Pomeroy said, that research would have to move in the direction also of ecohydrology in its next iteration so that modelling results could meet all the needs of predicting hydrological dynamics in complex physical and biological settings.

The goals of the next iteration of the existing research networks, he said, must be to advance our understanding and prediction of water and energy cycles in the western mountains and adjacent cold regions of Canada; to improve our ability to assess change and predictive uncertainty in mountain and cold regions headwaters that arise from changing climate and land use; and, finally and ultimately, to contribute to the assessment of the long-term sustainability of western and northern water resources.

In closing, Dr. Pomeroy noted that cold still matters in Canada. He said that the research undertaken by the IP3 and WC$^2$N networks created ideas and approaches whose time had to come – and that no one in either network should give up on the possibility of renewing this remarkable collaboration in the near future. He hoped that the next time this extraordinary group of people met, it would be in further service of the goal of understanding how and why cold matters to Canadians. Though it has yet to be scheduled, that meeting may be far more important to the future of Canada than anyone can imagine today.

# FURTHER READING

Arctic Council, eds. *Arctic Climate Impact Assessment*. New York: Cambridge University Press, 2005.

Beedle, Matthew J., et al. "Annual push moraines as climate proxy." *Geophysical Research Letters* 36, no. 20 (October 22, 2009): L20501, 5 pp. Abstract accessed May 22, 2012. www.agu.org/pubs/crossref/2009/2009GL039533.shtml.

Bolch, Tobias, Brian Menounos and Roger Wheate. "Landsat-based inventory of glaciers in western Canada, 1985–2005." *Remote Sensing of Environment* 114, no. 1 (January 15, 2010): 127–137. Abstract accessed May 22, 2012. http://is.gd/E4DaPr.

Cullen, Heidi. *The Weather of the Future: Heat Waves, Extreme Storms and Other Scenes from a Climate-Changed Planet*. New York: HarperCollins, 2010.

Daffern, Gillean. *Gillean Daffern's Kananaskis Country Trail Guide*. Vol. 2. Calgary: Rocky Mountain Books, 2011.

Ellis, Richard. *On Thin Ice: The Changing World of the Polar Bear*. New York: Alfred A. Knopf, 2009.

Groisman, Pavel, and Trevor Davies. "Snow Cover and the Climate System." In *Snow Ecology: An Interdisciplinary Examination of Snow-Covered Ecosystems*. Edited by H.G. Jones et al., c. 1. Cambridge, UK: Cambridge University Press, 2001.

Hansen, James, et al. "Climate change and trace gases." *Philosophical Transactions of the Royal Society A: Mathematical, Physical and Engineering Sciences* 365, no. 1856 (July 15, 2007): 1925–1954. Full text accessed May 28, 2012. http://rsta.royalsocietypublishing.org/content/365/1856/1925.full.

Jones, H.G., et al., eds. *Snow Ecology: An Interdisciplinary Examination of Snow-Covered Ecosystems*. Cambridge, UK: Cambridge University Press, 2001.

Lettenmaier, Dennis P. "Have we dropped the ball on water resources research?" *Journal of Water Resources Planning & Management* 134, no. 6

(November 2008): 491–492. Full text accessed May 30, 2012. http://
is.gd/741Fte.

Lovelock, James. *The Vanishing Face of Gaia: A Final Warning*. New York:
Basic Books, 2009.

Luckman, Brian, and Trudy Kavanagh. "Impact of climate fluctuations
on mountain environments in the Canadian Rockies." *Ambio* 29, no. 7
(November 2000): 371–380.

Orlove, Benjamin S., Ellen Wiegandt and Brian H. Luckman, eds.
*Darkening Peaks: Glacier Retreat, Science and Society*. Berkeley:
University of California Press, 2008.

Phare, Merrell-Ann S. *Denying the Source: The Crisis of First Nations Water
Rights*. Calgary: Rocky Mountain Books, 2009.

Pollack, H.N. *A World Without Ice*. New York: Avery, 2010.

Pomeroy, John, and Eric Brun. "Physical Properties of Snow." In *Snow
Ecology: An Interdisciplinary Examination of Snow-Covered Ecosystems*,
edited by H.G. Jones et al., c. 2. Cambridge, UK: Cambridge University
Press, 2001.

Sandford, Robert W. *Columbia Icefield: A Centennial Celebration*. Calgary:
Blackbird Naturgraphics, 1998.

———. *Ecology & Wonder in the Canadian Rocky Mountain Parks World
Heritage Site*. Edmonton: AU Press, 2010.

———. *Restoring the Flow: Confronting the World's Water Woes*. Calgary:
Rocky Mountain Books, 2009.

———. *Water, Weather and the Mountain West*. Calgary: Rocky Mountain
Books, 2007.

Sandford, Robert W., and Merrell-Ann S. Phare. *Ethical Water: Learning To
Value What Matters Most*. Calgary: Rocky Mountain Books, 2011.

Smith, Laurence C. *The World in 2050: Four Forces Shaping Civilization's
Northern Future*. New York: Dutton, 2010.

Struzik, Ed. *The Big Thaw: Travels in the Melting North*. Mississauga, Ont.:
J. Wiley & Sons Canada, 2009.

Weaver, Andrew. *Keeping Our Cool: Canada in a Warming World*. Toronto:
Viking, 2008.

Wiebe, Rudy. *A Discovery of Strangers*. Toronto: Knopf Canada, 1994.

———. *Playing Dead: A Contemplation Concerning the Arctic*. Edmonton:
NeWest Press, 1989.

# INDEX

*Ethical Water*
*Learning to Value What Matters Most*
An RMB Manifesto
by Robert William Sandford
by Merrell-Ann Phare
ISBN 9781926855707
4.75 x 7 inches
168 pages
hardcover
$16.95

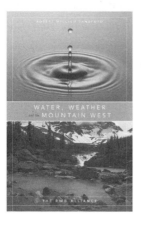

*Restoring the Flow:*
*Confronting the World's Water Woes*
by Robert William Sandford
ISBN 9781897522523
5.5 x 8.5 inches
304 pages
paperback
$24.95

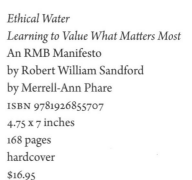

*Water, Weather and the Mountain West*
The RMB Alliance
by Robert William Sandford
ISBN 9781894765930
5.5 x 8.5 inches
208 pages
paperback
$19.95

# ABOUT THE AUTHOR

Robert Sandford is the EPCOR Chair of the Canadian Partnership Initiative in support of the United Nations "Water for Life" Decade and also sits on the Advisory Committee for the prestigious Rosenberg International Forum on Water Policy. He is a director of the Western Watersheds Climate Research Collaborative; an associate of the Centre for Hydrology at the University of Saskatchewan; and a fellow of the Biogeoscience Institute at the University of Calgary. As well, he sits on the advisory board of Living Lakes Canada and is co-chair of the Forum for Leadership on Water and a member of the Advisory Panel for the RBC Blue Water Project. In 2011 he was invited to be an advisor on water issues by the InterAction Council, a global public policy think tank composed of more than 20 former national leaders, including Jean Chrétien, Bill Clinton and Vicente Fox.

Robert is the author of some 20 books on the history and heritage of the Canadian mountain West, including *Ecology and Wonder in the Canadian Rocky Mountain Parks World Heritage Site* (AU Press, 2010), *Water, Weather and the Mountain West* (RMB, 2007), *Restoring the Flow: Confronting the World's Water Woes* (RMB, 2009) and *Ethical Water: Learning to Value What Matters Most* (RMB, 2011). He lives in Canmore, Alberta.